Nylon's Nativity

Dar...
to f...

Das ...tube onto the lab floor tiles.

Another synthesis then, but what?

Poly poly – not polyester... polyamide:

Eureka Polly

And thus was nylon – small 'n' born.

Disclaimer

To the best of the author's knowledge there has been no infringement of copyright or plagiarism. The story is based on a personal recollection of events over a period of some thirty five years – plus reference to basic factual information generally available, so more than a few mistakes may have crept in. If I have misrepresented people and events or failed to credit any source appropriately I am truly sorry.

Man Made Magic

When science meets fashion

by
Ronnie Price

Paperback ISBN 978-1-907685-64-4

ePub ISBN   978-1-907685-65-1

Mobipocket/Kindle ISBN 978-1-907685-66-8

Published in the UK by MX Publishing

335 Princess Park Manor, Royal Drive, London, N11 3GX

www.mxpublishing.co.uk

Cover Design by Martin Chiles

Dedication

When I first had the idea for this little book I discussed it with Nancy Richards who applauded it. We felt that I could do the basic stuff and Nancy would contribute anecdotes about the host of characters she had known during her long career in fashion.

I sent her some of the early text but then sadly she died.

It would have been a much funnier book if she had been involved.

And to Joe Scott who taught me so much about life; and Douglas Dickinson, Mike Page, Peter Owens and Bob Menzies from whome I learnt much about business.

Acknowledgements

First to the family: wife Sue who did the tricky stuff with website research; and daughter Jocelyn who with her husband Martin – both professionals in the graphic design business – did a superb job in making my scribble look presentable.

Warm thanks also to friends and former colleagues over many years who have provided pictures, anecdotes, and encouragement. In particular Mike Page and Phil Sharpe who went to a great deal of trouble in preparing material. David Buck, Peter Byrom and Norman MacCarthur sent helpful messages and I really want to thank the whole gang of past workmates for good times together. It was good to hear from Douglas Dickinson, my first boss at Courtaulds now retired in Australia, who sent me topical ryhmes from the rayon era.

And a final thanks to Steve Emecz my publisher for his guidance and support. This is our fourth book together ... and plans for more to come.

Contents

foreword by Brian Hamilton

The nineteenth century in the UK saw the gathering momentum of the industrial revolution. Powered by steam, electricity and inventive ingenuity, the foundations were laid for massive economic growth at home and imperial expansion overseas. 'North of Watford' a wide range of industries were established: engineering, mining, shipbuilding and, of course, textiles... cotton spinning and weaving in Lancashire, wool in Yorkshire, knitting in the Midlands and the Scottish Borders, and dyeing and finishing wherever there was water. Manufacturing businesses, often family businesses, sprang up led by imaginative entrepreneurs.

The first and second world wars gave added impetus to technical innovation, but commercially times were changing, not only for textiles but for manufacturing industry generally. Low labour cost countries, supported by modern machinery underpinning product quality, were increasingly competitive in what were becoming global markets.

Then, in the 1960's and 70's, UK textiles underwent a second industrial revolution stemming from two things: a) the accelerating development of man-made fibres and b) the formation of large 'vertical' textile groups.

The first commercially marketed man-made fibre had been 'rayon', made by extruding regenerated cellulose obtained mainly from wood pulp. Between the wars Courtaulds – originally Hugenot refugees – had a virtual monopoly of UK rayon production but after the war rayon was joined by wholly synthetic fibres such as nylon, polyester and the acrylics. The late 50's and early 60's saw the commercial launch of a wide range of new fibres with tailor-made properties which have since revolutionised textile products worldwide.

This man-made fibres revolution of the 60's and 70's provided the catalyst for the formation of a number of large textile groups in the UK. Prime instigators of this were Courtaulds, producer of 'rayons' but now also producing nylon, and ICI who had more recently entered man-made fibre manufacture through nylon and polyester. At this time the UK textile industry was structured 'horizontally' with fibres sold to spinners who made yarns, yarns sold to weavers and knitters who made fabric and fabrics sold to garment makers and household-textile manufacturers... with commercial interfaces and profit-taking at each successive stage. This was highly inefficient both operationally and financially.

In order to ensure market continuity for their fibre production, Courtaulds had already made moves to acquire textile manufacture downstream in particular in spinning. ICI, fearing such verticalisation by Courtaulds could jeopardise markets for their own fibres, in 1961 ICI made a bid for Courtaulds. Not surprisingly it was contested and battle was joined

In the event ICI's bid for Courtaulds was rejected, the main architect for the defence being Courtaulds' Frank Kearton, subsequently Lord Kearton who later became Chairman of the British Oil Corporation. It is rumoured that Courtaulds held a thanksgiving service in St George's Hanover Square, adjacent to their Head Office, to celebrate their deliverance from ICI – but one doubts if this is true. What certainly is true, however, was the continuing bitter rivalry between the two companies.

Courtaulds embarked on a major programme of downstream acquisition covering spinning, weaving, knitting, dyeing and finishing and garment manufacture. They also invested heavily in new state-of-the-art machinery, including several new weaving plants. This was matched blow for blow by Joe Hyman's recently formed Viyella International group, funded by ICI. Joe Hyman had a framed ICI cheque in his office for something like £12m, not a huge amount in today's money but a massive facility in the early 60's. Both Courtaulds and Viyella grew rapidly and others such as Coats Paton, Tootal and Nottingham Manufacturing joined the fray. When I joined Viyella in January '62 it employed around 1,800. Eighteen months later it employed something over 18,000, almost wholly by ICI funded acquisitions. And in the early 70's, on ICI's initiative, Viyella was merged with Carrington and Dewhurst to form Carrington Viyella, employing around 35,000 including 5,000 overseas.

I was a Carrington Viyella main board director. Ronnie Price was a senior marketing manager with ICI Fibres.

His book covers exciting times.

Brian Hamilton,
BSc.,MSc,.CText.,FTI.,FCMI,.Fins D.

introduction

Where science meets fashion

Why so magical? Well nylon, when it made its dramatic appearance in stockings in 1940, did seem like magic to women who had been accepting stockings which did not fit very well and did not wear very well.

Today man-made fibres, synthetics, don't arouse any excitement – few people would be interested in what fibre has been used in the clothes they are wearing, unless it is some expensive luxury such as cashmere, or silk; otherwise cotton or wool is assumed.

Just to clarify the term fashion we don't mean only size 6 models slinking down a catwalk although that's part of the clothing industry. New styles adapted from the Paris designer catwalk on sale in high street retailers in two shakes of a tail. Big business. A form of textile osmosis. So when we speak of fashion we mean not only the latest creations from big-name designers but essentially the clothes that trend setters amongst the public are wearing. And then for a time what a lot people are wearing until the cycle starts again with another significant fashion change – skirt lengths up or down, trousers skinny or baggy. This was very much the model in the past decades we are focusing on, although today a more eclectic mood reigns and to an extent anything goes.

A feature of the textile industry is its multi-stage pipeline: spinning, weaving/knitting, finishing, garment making, wholesale through to retail finally reaching the consumer. In the case of synthetics at one

11

end of this long pipeline from raw material to the consumer, is a complex chemical industry; and at the other an equally diverse textile and fashion business: science at one end; skill and creativity at the other.

During the twenties and thirties a new consumer market was developing with a growing demand for more attractive, reasonably priced clothes especially for women. This was partly the result of post Great War reaction to austerity but also the new force of feminism, and overall social change. It was also helped along and encouraged by the Hollywood films of the period which linked a more glamorous life-style with attractive clothing. Rayon spearheaded the response in this first phase providing fabrics which fed this wider fashion market in the twenties and thirties through to the second world war. The new war brought another development in the changing role of women in society but not much in the way of fashion if you exclude WAAF uniforms.

Then It Was Nylon making its debut in the early forties which was the new mover'n shaker in succeeding decades. As the first synthetic fibre to go on the market, nylon became the icon for a new major industry just about to burst onto the textile and fashion world.

It was an uncanny repeat of the twenties and thirties with early forces of feminism and social change; however in the sixties the peaceful revolution was much more of a fundamental game-changer.

Nylon's first role was in wartime for parachutes but then in New York in 1940 with a major launch the magic suddenly became apparent in stockings. As fine as if not finer than pure silk, much finer than rayon, and yet with much longer life than both, Nylon played its part in glamorizing legs in the early post-war era – somehow synchronizing with the Hollywood 'Technicolor' musicals in which lovely legs were a delightful main feature. Come back Betty Grable, wartime pin-up, all is forgiven…And Ginger Rogers 'who did everything Fred Astaire did but in high heels and backwards'.

Apart from the attraction of nylon as a new 'fashion' fibre it also related strongly to the trend towards convenience. Housewives were no longer prepared to be slaves in the home; war work had shown them a new world. Washing machines, tumble dryers and steam irons were their aspirational targets.

But nylon promised more and was becoming available: total easy care…non-iron, stain resist, ultra quick drying, low creasing. This was a brave new world, a part of the post-war promise and the consumer was ready to buy into it. Disappointment eventually set in and more realistic expectations emerged. Still, some of the initial magic – perhaps the consumer dream, remained and it set the stage for some twenty years of synthetic fibres in the textile and fashion world.

By the mid seventies, for reasons we shall consider, branded synthetics were in decline and the role of man-made fibres was changing with generic identities-nylon, polyester, acrylics, replacing trademarks, as just another type of clothing

Nevertheless for those of us who were privileged to play some active part in this era when the branded synthetics' rocket was shooting into the stratosphere, it was a very exciting and vibrant time. The new wonder fibres were not only practical but also fun, which is how we promoted them. And we had fun doing it in those halcyon days of the sixties.

Before looking at the main fibre types in more detail it might be helpful to give an idea of the basic requirements of a textile fibre.

In the case of natural fibres a yarn for knitting or weaving is made by twisting together individual fibres of cotton, or wool and drawing them into yarn. In order to be able to do this the individual fibres have to be long enough: cotton, wool, flax, and some other specialist hair and vegetable fibres, meet this criterion. Silk is a special case because it is extruded by the silk worm in long continuous 'filaments' which made it the choice for fine stockings and fine silks in the pre-synthetics' era.

Similarly to create man-made fibres the chemicals involved must have the capacity to form long chains of molecules which adhere together to provide strength, stretchabilty and the necessary characteristics for textile use. In synthetics these are known as polymers.

These long-chains polymers are converted into yarns or fibres by melting the polymer chips and extruding – sqirting, the liquid produced through 'spinnerets' heads with small holes – in the form of continuous filaments – emulating silk. Thick ropes of filaments can be combined to be cut into staple lengths of approximately the same dimensions as cotton or wool.

A major feature of synthetic fibre chemistry was the ability to use different routes to create a variety of fibre types.

Initially the goal was to replicate natural fibres. For example rayon was known as 'artificial silk'. However with the true synthetics, and the range of possibilities, the game changed and a fresh target was to improve on natural fibres in a number of ways. We shall examine the success of that approach.

How it all began with rayon – the first artificial silk

The expression 'man-made fibres' is an umbrella for various categories with pure synthetics and re-generated fibres being the two main types involved. Synthetics are the product of chemical processes, mainly based on oil derivatives; we can look at the detail later. In the case of re-generated fibres a form of cellulose in the form of wood pulp, is dissolved by chemical process and then reproduced in fibre form by squirting it out through spinnerets. The end product is still chemically cellulose, the same as the original source material. It has simply had a change of state into a new form, as a textile fibre.

At the end of the 19th century there was ambition to find an alternative to silk: to create an 'artificial silk'.

One of the earliest attempts in France was based on

nitro-cellulose a highly flammable and volatile material; a version was also used to make the film used in cinema projectors and resulted in some alarming fire outbreaks which for a while threatened the future of the cinema in its early 'picture house' days.

The artificial silk story did not really take off until the development of viscose rayon. The manufacturing process was relatively straight-forward and volume production was quickly established. Today there would have been environmental consternation because of the fumes given off by the chemical activity...and it was quite a 'pong'! The partial answer was exceptionally tall factory chimneys and these became landmarks of the factories of Courtaulds – the main company involved in rayon making – around Britain. At the Coventry factory there was a king size chimney. The distinctive smell led to a popular local rhyme.

Courtaulds built a chimney,
It wasn't meant for smoke,
It takes the smell from Foleshill
And it puts it down in Stoke.

They were rather interesting architectural achievements in their own right as an image of that mid industrial age.

Courtaulds was a Huguenot family which like many others had moved to England in the 18th century to avoid the persecution of protestants in France. There was a strong tradition among the Huguenots of silk processing and weaving and they brought these crafts to England, although initially the Courtaulds family became silversmiths in London and – only later – leading weavers of mourning crepe with factories in several small Essex towns.

At that time funeral crepe was big business with women taking seriously the matter of dressing in an appropriate manner for what was an important community occcasion. Moreover the general style of clothing in the period used a considerable amount of fabric for women's dresses and it was good business with some 3000 workers being employed at the peak of customer demand. But it was to decline as fashion and social mores changed.

The Courtaulds Family were enterprising and with their silk background spotted at an early stage the attraction of an 'artificial silk'. A new non-family director, brought in to restore the business after mourning crepe declined in popularity, purchased in 1904 the patent to manufacture viscose rayon in Britain for £25000. A bit of a gamble but it paid off. At first there were problems with process but with their textile background Courtaulds were able to overcome these and after just a few years had a very successful and expanding business.

The first factory was set up in Coventry in 1904. They may have chosen Coventry because several Huguenot families had set up silk weaving in that city. They expanded rapidly through the twenties and thirties with factories in a number of British towns including Preston, Nuneaton, Wolverhampton and Flint in Wales where a former German factory was taken over during the first world war. There were also plants in Northen Ireland, France, Germany and America. They set up an American company early on in 1909 taking over the original American patent for covering the same process as the one they were using and in England creating the American Viscose Company. Ultimately the American plants were lost during the war when they were taken over as payments for arms under the Lend Lease programme. This included 50 clapped out first world war destroyers, but never mind they were most welcome at the time. As it worked out the American business after some very profitable years was by late thirties struggling so perhaps it wasn't too great a loss.

There were other forms of rayon and other producers but Courtaulds was the dominant force in the market not only in Britain but internationally at that time. They were pretty adept marketeers and post World-War Two introduced a 'Tested Quality' labelling scheme which although never really successful may have paved the way for the subsequent BNS 'Bri-nylon' programme. At that time the concept of committing large advertising budgets to persuade consumers go into shops to look for and ask for brands by name, had not been born.

Cuprammonium was one variant of rayon with some good characteristics but it was more expensive than viscose and never became a serious contender apart, mysteriously, for sleeve linings in expensive tailored suits!

British Celanese the Main Competitor produced a type of rayon known as Cellulose acetate. Its character made it the choice for underwear and 'soft drape' dress fabrics but it was not suitable for the big stocking market and it was more expensive to produce.

A little verse of the period underlines its place in the lingerie sector.

When daughters of Vicars wore nondescript knickers,
They often got left in the lurch
But now they spend Sundays in Celanese undies,
You can't get a seat in the church

Reputedly the most profitable part of the cellulose Acetate operation was effective recovery of acetone during the process! From a marketing perspective 'Celanese' was one of the early successful uses of fibre branding with the 'Celanese' name being used on garments. A modified version tri-acetate was a step nearer to the pure synthetics with its facility to be 'heat-set' and it played a part in the fashion for pleated skirts and dresses in silk-like fabrics under the brand 'Tricel'.

Courtaulds expanded production of its own cellulose acetate and then in a decisive business strike consolidated its position by taking-over British Celanese in the sixties.

<u>*There is*</u> no doubt that rayon played a major role in the fashion boom of the twenties and thirties. It was helped by the impetus of dispelling the post war gloom but unless a cheap mass market fibre had been available it could not have happened on such a scale. It played its part in dressing the shop girls, typists and mill lasses for the jazz age. Yes the high society flapper girls were still doing it in silk, high kicking the Charleston in the Mayfair night spots; but up and down the land in Britain small bands and orchestras were providing the music for the Saturday night dance in modest local halls. And the short skirt fashion could not have worked without the rayon 'artificial silk' stocking which meant

that girls ,who could not possibly afford pure silk could now show off their legs.

During the thirties there was a considerable amount of technical progress with rayon: a dull 'Delustra' yarn-option; spun dyed yarn with the colour introduced during the manufacturing process; and staple fibre in which the rayon is cut into short staple lengths as part of the spinning process designed to be made into yarns like cotton or wool. It was also developed for industrial use and Courtaulds 'Tenasco' high-tenacity rayon fitted well into the burgeoning car and truck market. This was serious high volume business at that time and of course growing although interrupted by the war apart from military use.

Yet Behind the rayon success story there was in the scientific world continuous research to find the Holy Grail…a truly synthetic polymer which could be used to make a 'miracle' textile. Although chronologically it came much later, the 'man in the white suit' film portrayed this goal. The ghost of Alec Guinness receiving the opprobrium of both mill owners and workers who saw his everlasting fabric destroying their livelihoods.

It would be some time before the various strategies to meet this target would become viable; and some of them never would. The idea of polypeptide long chain polymers replicating the chemistry of natural silk has not made it so far.

The story of rayon did not end with dramatic appearance of the synthetics. Modified forms of rayon were introduced into the textile trade: examples are polymeric and hollow fibres. Some forms were stronger than both cotton and standard rayon, especially when wet which was always a problem for cellulose based fibres. Today modified rayons are still playing a role in blended fabrics for the fashion industry.

Nylon the game changer

Walter Hume Carothers was recruited, somewhat reluctantly on his part, to take a leading role in Du Pont's newly established programme of research into organic chemistry polymers. We rather like to think that major inventions come about from some gifted individual getting a bright idea while working in his garden shed: 'Eureka I have it!' Or by chance in the lab, when the scientist is actually looking for something else. Well perhaps that is sometimes the case: is Fleming's discovery of penicillin an example?

The reality is that in the main progress comes from a strategy backed with resources, yet perhaps involving some talented individual knowing exactly what the objective is and adding a flash of genius to the structured research.

Du Pont was one of the biggest chemical companies in the world and when in the late 20s they instituted a programme of 'pure scientific research' it laid the base for the work on polymer science which led ultimately to the discovery of nylon. It may be useful to explain something about polymers. In our everyday language they are formations of chemical molecules which have the capacity to stick together and produce 'long chains' – suitable for forming fibres.

Carothers had resources, skilled assistants, and excellent equipment at Du Pont and he had the satisfaction of his achievement; but there is a theory that he might have enjoyed more personal happiness and avoided his ultimate suicide if he had remained in the scholarly atmosphere at Harvard university – although it is perhaps questionable whether academia cannot be every bit as competitive as the big bad outside world.

As it was he took the job and was successful in a field of research which fascinated him. It was not all plain sailing though even after he made the breakthrough with polymers. The first promising results came with a form of polyester. After a delays and some dispute, while he negotiated with his employers, he presented his research to an important scientific congress in 1931.

There was popular acclaim for this promise of a new material from a revolutionary synthetic process. It was the era of new products-plastics such as Bakelite – and especially in the US these 'inventions caught the imagination of consumers linking with the force of upward aspirations within the new force for social change. American advertising reflected these aspirations.

According to some contemporary Carothers went

through an uncomfortable period with Du Pont with some disagreement about the role of fundamental research. He still had an academic bias and wanted to publish his results openly in scientific papers but the company now had to have a more acute eye on the commercial potential and their expectations had moved more concretely in that direction. They were understandably wary about giving too much away to potential competitors. They knew that around the world several companies were already devoting resources to polymer research.

A significant step came when Carothers switched his attention to polyamides because of their higher melting point compared with the polyesters used so far in his experiments. In fact it was not until Rex Whinfield the talented British scientist did critical work on ways to spin polyester that it became feasible to use its chemistry as a basis for a textile fibre material.

Carother's change of focus to polyamide was successful and led to the development of polymers with the characteristics required for textile yarns.

His contribution to polymer science was considerable and his death in 1937, before the commercialization of his invention nylon, was a sad blow.

It seems that although he was first and foremost a scientist and ostensibly a little naïve, he saw the potential of nylon as an apparel fabric while some of his Du Pont colleagues were still thinking of it as essentially for industrial use. So it is a quirk of fate that he missed its consumer launch in the form of stockings in 1940 in America. The impact was such that before long stockings became known as nylons and the demise of rayon stockings became inevitable. But it was not until the early post-war period that nylon stockings became available in Britain. That is unless you count the nylon stockings used by enterprising GIs as the ultimate seduction tool. English girls preferred them even to Hershey chocolate bars or genuine American chewing gum.

An interesting flash-back to the twenties shows an historic replay in the stocking market. The 'flapper' skirts in that earlier time were encouraged by the advent of the fully fashioned rayon 'artificial silk' stocking. Forty years on nylon tights in due course made the mini-skirt acceptable for everyday wear: the alternative of showing suspenders to keep up stockings may have appealed to a few prurient males but girls would not have felt comfortable going up escalators on the London underground. Tights have never been popular with men earning the soubriquet 'passion killers', the term used pejoratively during the war to describe the stout bloomers issued to the women's services One of the leading hosiery manufacturers tried to spice up tights, using the American term panty hose and introducing a simulated knicker with pastel frills. The first nylons were fully fashioned-knitted flat and then seamed together giving that distinctive seam down the back of the legs look still rather admired by discerning men-folk of a certain age. The advances in knitting technology in the late fifties introduced the 'seamless' stocking. These stockings, and later tights, were knitted in the form of tubes and then heat-set in the shape of legs on special formers. One of nylon's important properties was its thermo-plastic nature which meant that it could be 'permanently' set into shape.

Arguably Du Pont missed a trick when they failed to register 'nylon' as a trade mark. It must surely have occurred to them so presumably there were legal trade mark reasons why they could not. Thus nylon became a generic name and this created some interesting challenges and opportunities to nylon producers as the industry developed and patents ran out. Subsequently Du Pont have been quick off the blocks in branding their textile fibres-for example 'Lycra', 'Orlon' and modified forms of nylon.

After the phenomenal success of nylon stockings the objective became the expansion of nylon into other apparel sectors. At first the enthusiasm for this new wonder fibre encouraged some unwise moves into garments where the characteristics of nylon were by no means appropriate. Shirts made from parachute type fabric which had been designed to prevent air passing through did not give nylon a good name in the comfort stakes.

By the Time nylon was becoming more readily available in the UK, the home producer British Nylon Spinners (the name says it all) were aware of the need for a more systematic approach and a use development organization – supported by a technical development department – was set up at the new Pontypool plant in Wales. Pontypool was a greenfield site but by 1964 it employed 5700 people. In 1963 BNS was the biggest nylon producer in Europe and responsible for 10% of the total world nylon output. Pontypool held another record: the clubhouse held the largest dance hall in Wales which could hold six hundred dancers!

The nylon yarn was spun at Pontypool, and later other plants, but the basic polymer feedstock came from the ICI plant at Billingham, later from Wilton.

The BNS story is interesting because it was a brand new company set up within an impressively short timescale, to develop, manufacture and market a totally new product. The new organization...management, workforce, sales-team – all new to this business – was soon working effectively. FC Bagnall, 'bags', was the managing director who provided strong creative leadership. He later became a member of the ICI main board.

To be fair there was good outside support. BNS was operationally an independent company but it was jointly owned by ICI and Courtaulds. During the 40s ICI had taken a Du Pont patent to manufacture nylon for wartime strategic use and since they lacked at that time specific textile skills and experience they entered into partnership with Courtaulds to provide textile know-how. This continued for some time into the post war years with first nylon production at Courtaulds Coventry headquarters.

BNS, possibly because of its youth, was in some ways more adventurous than its parents and it raced away with a brave approach to marketing. The development activities rapidly took nylon into different products. An interesting example is the shirt sector. It was recognized that pushing the wrong fabrics into the shirt market could damage the image of nylon and a mechanism was required to control this. They were also well aware of Du Pont's failure to brand their nylon.

These factors led to a major programme for the

branding of BNS nylon as Bri-nylon. It was a unique system which monitored the quality of the fabric and garments made from BNS nylon. It used a deferred trademark which meant that the Bri-nylon brand could be used only on a final garment – such as a shirt – if the garment make-up was satisfactory as well as the fabric used. The garment maker had to be registered as an approved user of the Bri-nylon name. So even if BNS yarn had been used the garment could not carry the Bri-nylon mark unless BNS approved. Around one thousand companies signed up as registered Bri-nylon users.

This programme was supported by advertising and promotion right through to the retailer and consumer. This was something revolutionary in textiles and represented a major move in modern marketing concepts. Even today it still represents a very sophisticated level of marketing technique. It involved working with all the different trade levels helping them to adapt to a new textile yarn in fabric production and garment making, as well as helping each trade sector to promote its product. Initially there was an interesting marketing dilemma. An optional name was Bri-lon. There was a view that it sounded friendlier and softer than Bri-nylon and in fact created an entirely new image separate from the generic term nylon with its sometimes negative associations in comfort characteristics. The opposite view was that nylon was too strong an identity to step away from. So both names were deployed at first with Bri-lon being reserved for knitwear. Eventually the costs of promoting two brands dictated the choice of one only and Bri-nylon won the toss.

The main architect of this unique programme was

commercial director Howard Morris and later marketing manager Eric Sharp (who later became Lord Sharp and chairman of Cable and Wireless).

We can take men's shirts as an example of product development. The initial effort was to get the right fabric. There were various approaches: woven fabrics based on textured filament yarns or schappe spun. Filament yarns are those which are in a continuous thread like silk for stockings; and schappe is when the fibre is chopped up and put through a process of spinning similar to that used for cotton. The process is sometimes known as 'spun silk'.

However it turned out that the one which worked in the volume market was warp-knitted fabric. Warp knitting was perfect for nylon. New knitting machines could produce a woven type fabric at high speed and the smoothness and high tenacity and elasticity made nylon the ideal partner. Moreover the warp-knit fabric could be designed to make it breathable and comfortable by using an open structure fabric.

Having the right fabric was the start point but it was then necessary to ensure that the right techniques were used to make up the garment. BNS recruited making-up specialists to advise on non-shrink sewing threads, tensions, interlinings and collar fusing, before a shirt-maker submitted a sample for testing.

The Bri-nylon shirt programme was very successful and at one stage had around 40% of the British shirt market including special counters in Marks & Spencer stores. Of course new fabric types and patterns were required to maintain share in the fashion market and BNS responded with new products such as melt-spun coloured yarns and finer deniers (yarn thickness) for the 'Tricopress' brand. The 'Tricopress' derivative was based on knitting machines with a finer gauge to give exclusivity, linked to a clever funding programme for promotion and advertising in which BNS matched a contribution from the knitter.

Innovation and development also played a part in other sectors – lingerie, bras and corsetry, swimwear and dress fabrics. Yarns for stockings were improved to suit new high speed circular knitting machines; and new types such as 'Tendrelle' introduced softer finish stretch yarns for the hosiery sector.

One of the Challenges in women's wear was to gain the support of fashion designers and influential figures in the fashion world. A small team was created to concentrate on this. They attended fashion shows in London, Paris, Rome and other international centres to build relationships in this somewhat exotic branch of the textile industry. Exotic it may have seemed but behind the frothy façade the people involved were every bit as hard headed as the traditional Lancashire mill owner. To win their support it was necessary to persuade them to try nylon fabrics. This involved working with selected weavers, and knitters, to develop fabrics which would suit the couture world and led to high-fashion garments incorporating nylon appearing in designer collections.

This was part of the strategy adopted at that time of establishing a positive reputation for a brand right at the pinnacle of the marketing Christmas tree to give it an image with which eventually to penetrate the volume retail field. With a little marketing lubrication the use of nylon at designer level trickled down through fabric producers and women's wear garment makers into the popular market place. This was helped by the growing trend for leading fashion buyers from major retail stores to attend designer shows and to buy garments for their top-end fashion departments. They could use some of the garments purchased to inspire their 'tame' clothing suppliers to manufacture evolved styles for their more popular ranges. And fabrics containing nylon used in the collections were suitable for commercialization.

This technique is not unlike the well-known advertising stratagem of using personalities in advertisements, by association endorsing the product involved.

The 'household' or 'domestic' market used not to be regarded as part of the fashion scene but my how things change. Many respected fashion designers now have a finger in the new tasty domestic textile pie, Ralph Lauren being a good example This was not the case in the early days of nylon where household textiles tended to be the province of some big established specialist companies producing sheets, bedding, curtains, upholstery fabric and loose covers. Some major retailers had their own labels.

In fact nylon did not slot readily into this field. Warp-knitted brushed sheets made the biggest penetration but apart from price and their non-iron quick-dry washing facility they never really caught the public's approval in the middle market sector with a general negative reaction to comfort. In due course they were displaced by polyester/cotton blends. Of which more later. The nylon sheet saga showed how a product's image could be degraded if the product is not right; and if a retail price war forces quality cuts and consumer dissatisfaction.

On the Fringe of household textiles is the very large carpet business and once it was properly developed and established nylon played a significant role in this – and still does. The first break came in the States with a nylon specially developed by Du Pont for a new form of carpet manufacturing known as tufted carpets. Nylon and tufting, like nylon and warp-knitting was the proverbial marriage made in heaven. They could work together to make carpeting on a mass production basis that made traditional woven carpet making look rather antediluvian. Again the timing was good: brave young Americans were buying new homes and carpeting seemed the smart way to establish an up-market look. It was part of the new furniture design trend; and being nylon also promised the long life and easy-care which were part of the modern housewife's dream.

But despite evolutionary types of nylon fibre with a different cross-section (the shape of the fibre if you were to cut through it) the tufted nylon carpet did not really live up to consumer expectations. Somewhat surprisingly bearing in mind nylon's outstanding ability to recover from stretch and stress deformation, reasons why it is outstanding for stockings and tights, the carpets were prone to flattening giving a 'dead look'. More predictably, even though special anti-static finishes were applied at fibre or fabric stages, the carpets attracted and retained dust, dirt and stains. Another factor which did not help with the performance was again the inevitable cheapening of the product to gain competitive price advantage. If the manufacturer used wider stitch patterns the pile was not so firmly packed together. Less nylon yarn was required and production time reduced. Dollars were saved and this meant it was easier to compete in the lower end of the floor covering market. The first high quality tufted nylon carpets were pretty good but after a time tufted nylon carpets lost status. This was perhaps less so in America, but certainly in the UK the product fell to the bottom end relatively quickly in spite of determined efforts of entrepreneurs like Cyril Lord who established a major plant in Northern Ireland. These days with new blends and qualities it seems to have found a niche.

You could say that in Britain, nylon's bacon was to some extent saved by some good work by BNS/ICI, who took an alternative approach. They worked with traditional carpet yarn spinners and weavers of wilton and axminster to design a Bri-nylon fibre for blending with wool. The standard blend was 80% wool 20% nylon although some variants were used when searching for the optimum. Eventually 80/20 was settled on and it became very popular – and still is. This blend was very hardwearing and had a traditional wool woven carpet look. It was taken up by some of the top carpet weavers such as 'Brintons' and actively promoted. It also worked well in heavy duty contract use.

Although a less glamorous side of the nylon business its use for industrial purposes was a significant contributor to the volume of sales. By their nature industrial products tend to use a lot of fibre. In this field nylon's modulus of elasticity and high tenacity were particularly attractive. It was – and is – used in a very wide range of outlets.

It took over from rayon for tyre-cord fabric but subsequently gave way to steel cords.

It proved to be good base fabric for coating and was deployed for tarpaulins and protective clothing. Ropes – including massive hawsers – and cordage, including fishing nets were also important outlets. Yachtsmen racing competitively found it particularly good for spinnakers able to balloon and enclose considerable quantities of wind pushing up speed. Nylon was used for hose reinforcement, and inflatable buildings to act as marquees or temporary structures and it pioneered a new form of lightweight luggage.

In some outlets it has now been replaced with other synthetics, but it was the first synthetic to challenge traditional fibres like cotton flax and hemp in this field.

For some years the national markets in Europe and elsewhere for nylon remained discrete protected by patents or licences. BNS perhaps more than any other European producer took a major step – with the Brinylon branding programme – to protect their business when patents finally expired. The patents were extended by the introduction of new technology but eventually protection would be gone.

In mainland Europe there was a major competitor to Nylon 66 (the type of nylon invented by Carothers and made by BNS and Du Pont). Dr Slack director of research at I.G.Farben, the German chemical giant had discovered an alternative to the Du Pont patents. He and his team fairly quickly developed a new type of nylon. This was nylon 6 which became known as 'perlon'. It was taken up by several chemical fibre producers including AKZO, Holland; Snia Viscosa,

Italy; Rhodiatoce, Italy; and Bayer, Germany.

Perlon was not as good as Nylon 66 in some aspects of technical performance and was generally marketed at a discount but it was a major player in continental Europe. Eventually BNS invaded Europe by setting up its own plant in Oestringen, Germany – a very brave move at that time.

There were interesting power plays in Britain. ICI mounted a major bid to take over Courtaulds – described as a merger – and to help develop a stronger UK textile market. Quite early on it was realized that to some extent the future of the fibre producers was linked to the strength of their customers: spinners, fabric producers and clothing manufacturers. Courtaulds were already strong in this down – stream business area and since ICI and Courtaulds were already partners in British Nylon Spinners this seemed a good idea to ICI. But Courtaulds wasn't playing. A fierce battle raged in the financial press with counter claims and offers to shareholders from both sides. Sir Frank Kearton led a spirited defence for Courtaulds; he was a highly regarded scientist from Courtaulds famous No.2 lab, responsible for a major break through in acetate rayon processing. The ICI chairman was Sir Paul Chambers, said to be the first non-scientist to occupy that top ICI spot.

ICI achieved 38% of Courtaulds shares but eventually decided to withdraw. A deal was struck, the basic arrangement being that ICI would relinquish its shares in Courtaulds, and take over the whole of British Nylon Spinners. It was also agreed that Courtaulds could develop and market its own form of nylon. It had clearly been working on this for some time because it was not too long before 'Celon'-nylon 6-appeared on the market.

The Dutch giant chemical company Akzo also set up a plant in Northern Ireland to manufacture its brand of nylon 6, 'Enkalon'.

Competition in then nylon market in Britain was definitely hotting up but British Nylon Spinners marketing strategy with Bri-nylon-and its strong product range payed off and neither 'Celon' nor 'Enkalon' proved to be serious market competitors. In fact they wisely concentrated on 'niches': 'Enkalon' went largely into tufted carpets and 'Celon' was used mainly in Courtaulds own warp-knitting plants.

So until the general demise of branded synthetic fibres, to be explored in a later chapter, Bri-nylon ruled the roost in Britain and made some impact in Europe through its German operation.

BNS was active from the beginning in export with Commonwealth countries being obvious target markets. Australia, New Zealand, South Africa and India were especially important. As the business grew local manufacturing was established: Fibremakers in Australia and S.A.N.S in S.Africa. In India the industrial uses of nylon were of special interest as the country began a programme of economic development.

The Nordic countries, especially Sweden, were important markets and had no local production.

Nylon led the drive by man-made fibres into the world-wide textile industry. In Britain the part played by BNS was distinctive and dynamic.

3

Polyester the new contender

Although originally Carothers had worked with polyester for his polymer research it had not provided a satisfactory result and he switched his efforts to polyamides leading ultimately to the discovery of nylon.

However in Britain there had been a serious interest in polyester in its own right. In the mid thirties Rex Whinfield a senior scientist working for a textile processing firm, Calico Printers Association, began research into the possibilities for synthetic fibres. He was almost certainly influenced by the published work of Carothers but pursued a different chemical route. His bosses at CPA felt that this work did not fit into the scope of their normal business scope which was dyeing, finishing and processing fabrics, a field in which they had a well established market and reputation. So it is to their credit and perspicacity, that they did in fact allow Whinfield to continue his research to develop a polymer to be used as a basis for a textile fibre. He got the full go-ahead in 1940 and identified surprisingly quickly the chemistry for a fibre-forming polymer based on polyester. The war effectively delayed development but by the mid-late 40s a programme for post war commercialization was established.

British Chemical Giant ICI had initially taken the view that its role in synthetic fibres would simply be the production of the polymer the raw material to be used for spinning, and not to undertake the actual textile spinning process itself. This was the pattern of the co-operation in the joint venture with Courtaulds to make nylon. However new forces were at work in ICI and textile fibre manufacture became a specific goal. Various possibilities including acrylics were being studied.

So Whinfield's 'Terylene' as it became known was of singular interest. An arrangement between ICI and CPA, partially brokered by the Ministry of Defence which saw strategic potential, was made in 1943. Then, post war CPA, transferred their patent rights worldwide to ICI – except for the USA which opened the door to Du Pont into polyester development and manufacture.

ICI quickly put resources into 'Terylene' development although there were many technical aspects to be resolved before a major move to market it could be made. A 'Terylene' council was set up by ICI to manage the project originally in Welwyn linked to the plastics division then to a purpose built headquarters in Harrogate.

It is interesting to look at the background to Rex Whinfield's discovery which took place in a small but well provided CPA research laboratory, Broad Oak in Accrington. He paid tribute to the support of a junior colleague, the Australian James Dickson and to the pioneering work of Carothers. In a subsequent article in a Textile Institute magazine he commented on the relative costs of the initial research and development work compared with commercial scale manufacture. The original lab work might have cost CPA less than £50,000 per annum based on Forties' values. To bring it to commercial manufacturing probably cost six hundred times that sum and that was before substantial marketing-including advertising-costs.

It needed a pretty big group with substantial resources to undertake this level of investment and so it is hardly surprising that the main players in manufacturing synthetic fibres were chemical giants, some with government support as in Italy. Ultimately this became a major problem. For too long – during the later years of fierce competition – fibre producing operations continued despite posting huge losses because their parent chemical groups could not let them die as would have happened if normal market forces had been allowed to apply; instead their parents had the resources to continue propping them up even when massive losses piled up.

ICI Now Had a foot in both camps: nylon through their half-share in British Nylon spinners; and their own polyester. But as it worked out there was little direct competition between the two types of fibre. Nylon because of its modulus – high stretch and recovery characteristics – was perfect for stockings and tights. Polyester had a different modulus which gave it good short term stretch but poor recovery if it was stretched too far. Stockings which bagged round the ankles were not appreciated by the ladies. On the other hand polyester's properties made it ideal for blending with natural fibres and at first this was its main potential.

In the continuous filament field polyester did score initially in net curtaining where its resistance to sunlight degradation made it very popular. At one time the semi-detached homes of British suburbia seemed swathed in Terylene net curtains. One may speculate about the millions of busybody eyes, on the lookout for neighbours' indiscretions, which operated behind this discreet screen. 'Coronation Street' terraces in their finest hour

The performance difference is summed up in the use of the two fibres in the world of sailing. Polyester is used for the main-sails because of its firmness and low stretch; whereas nylon is chosen for spinnakers where its high stretch character allows it to optimise the wind for racing speed. Polyester was better too for fabric to be used for men's ties. Its firmer character made a more solid fabric than nylon.

This difference was to play a part in perhaps the most important development of the polyester business – blending with other fibres especially cotton and wool. Such blends are standard today. Apart from carpets nylon never really became a force in blending with other fibres. Its modulus was not compatible with the major natural fibres-cotton for example – although it did work with viscose rayon in small blend levels.

To some extent there was indirect competition between nylon and polyester in the UK in the shirt sector. The Bri-nylon shirt was preferred if drip-dry was a main concern. There was some criticism on the grounds of comfort but the very high market share it enjoyed for several years offers another view on this. And it overcame competition from resin treated easy-care cottons despite heavy promotion for them.

But it was a very different story in the United States. Du Pont never tried seriously to develop warp-knit nylon shirts, possibly because their branded polyester 'Dacron' was seen as their major product for the shirting market in the form of cotton blends. The polyester/cotton blend shirt became a standard mass market product. There was disagreement about the relative percentages of polyester and cotton. The polyester fibre producers wanted higher levels of polyester: 65% polyester was an initial blend recommendation but market pressure forced them to accept 50/50 and even reverse blends, 65% cotton and only 33% polyester. Fibre producers to some extent controlled this situation by allowing only fabrics which met their specification to carry their brand on swing tickets and labels. The lower polyester blend protaganists argued that fabrics with a higher percentage of cotton were more comfortable and less prone to 'pilling'. The counter argument was that without sufficient polyester the easy-care performance and longer wearing ability would be compromised.

In the US Du Pont eventually adopted a more flexible policy, perhaps their much larger home market encouraging this approach; while in Britain ICI adopted a firmer policy in keeping with their general approach to place emphasis on quality standards whenever their 'Terylene' brand was used.

It is arguable that a softer regime in respect of blends could have encouraged a bigger UK market share for 'Terylene'/cotton shirts; although another significant factor in the success of polyester/cotton shirts in the US was the general availability of large home tumble-driers which provided the best way of exploiting the easy-care properties of the blend.

At that stage tumble driers were not standard in British homes, and still are not common today. The preferred method back then was drip-dry over the bath and Bri-nylon could not be bettered for that.

Men's outerwear-suits, jackets, and especially trousers – was an early outlet for 'Terylene' in the UK and advanced quickly. 55% 'Terylene'/45% wool became very popular and for good reason. It had the appearance of a fine quality worsted trouser but offered excellent pleat or crease retention while resisting unwanted wrinkling. Before long with suitable interlinings these trousers could also be machine washed.

A serious attempt was made by ICI to introduce a higher Terylene content blend for menswear. In addition to the standard 55% Terylene a further percentage of high bulk fibre was added, the high bulk fibre carried a price premim. It had the objectives of selling more Terylene, part at a higher price and producing a better handling cloth. It was branded as

Terylene plus T. It was not as successful as had been hoped, largely because of a world collapse in wool prices at the time it was launched.

After being launched in classic style in higher quality menswear Terylene soon established a strong position in the volume market through the chain stores, a growing force in the sixties and in multiple tailors, a menswear retail system unique to Britain. At Burtons, Colliers, Hepworths and several others you could order a suit on a Saturday afternoon and a cloth pattern book, pay a £5 deposit and collect it about a month later. Polyester staple fibre also made its mark in various blends, not only wool and cotton but in linen and some specialist fibres.

Although at first the shirt story was the most important one for polyester/cotton blend, it also became popular in rainwear, especially men's raincoats. The blend fabric was light and with an appropriate finish could be made adequately shower proof. With the right type of interlining and sewing thread it was also a machine washable garment, useful for travelling. Perhaps because of that facility these coats tended to be in lighter colours.

'Terylene'/worsted blend fabrics produced polyester's first success story in womenswear. This was the pleated skirt which virtually created a new fashion in leading chain stores and high street medium level fashion shops.

It Was In the mid sixties that one of the most intriguing stories in the textile and fashion industries took place. As with some other legends there was symbiosis in technology and serendipity in the timing.

There had been a growing interest in the idea of knitted fabrics for women and suitable circular-knit machinery was becoming available. The attraction was the speed of the knitting machine, simpler preparation of the yarn compared with weaving, and the promise of a new type of fabric to stimulate women's fashion.

The problem for the knitting industry was that although the right machines-circular double jersey-were coming on stream the yarns then available were not really adequate for high-speed knitting performance nor did they provide the easy-care which had become a factor in consumer preference.

For some years ICI had been attempting to create a yarn which would fit into the double-knit sector. The first efforts had not produced worthwhile results However 'Crimplene' the invention of technologist Mario Nava, who was working at British Depa Crepes, introduced new technology. It was taken up by ICI and proved to be a powerful game-changer. The process used setting temperature and 'false-twist' to give high bulk yarns which were steam set to make them stable and very suitable for a high -speed knitting operation on double-jersey machines. ICI introduced a new basic filament yarn known technically as 167 decitex /30 filaments as the feed-stock for the 'Crimplene' process. The latest knitting machines and the 'Crimplene' yarn went together like the proverbial horse and carriage although perhaps a more elegant fashion simile should apply. Another timely coincidence of technologies.

The potential attracted entrepreneurs and one such was Henry Knobil. He had been a textile technologist at Marks & Spencer and was therefore well positioned to keep abreast of technical innovation and market opportunity. He left M&S in 1964 and set up a double-knit jersey company Textured Jersey. It rode the Crimplene tidal wave and contributed to its performance by introducing jet dyeing, finer gauge fabric and heat-transfer printing thus broadening the design possibilities for the fashion market.

'Crimplene' became a 'must-have'. The fabrics could be produced in a range of weights and patterns and they could be fabric dyed into attractive fashion shades or knitted from dyed yarns. Jacquard – computer designed – patterned fabrics gave garment stylists fresh opportunities. Using specially adapted techniques the fabrics could readily be made-up into the latest styles. This suited the new volume market built around high-street chain stores. 'Crimplene' clothes delivered: drip-dry, crease resist, anti-crushability and good draping aesthetics. They were also comfortable to wear and perfect for travel. It was very well promoted by ICI and became the market leader. Other fibre-producers brought out their own versions in due course but never matched Crimplene, partly because ICI had the initiative with the technology but also because of superior commercial strategy and clever marketing. 'Crimplene' output at the end of 1968 was reckoned to be more than fifteen million pounds, impressively almost double that of total American production of textured polyester.

ICI did not sell the 'Crimplene' yarn direct to the knitting industry. They licensed a limited number of firms, traditionally called 'throwsters' to produce it from ICI flat yarn and sell it under strict technical control to ensure that consumer expectation was fulfilled. A 'Crimplene' licence was highly prized, not quite a licence to print money but perhaps not so far off! AN early example of quantitive easing. Initially there were six licensees but later three overseas companies were included. Not only was the textured yarn tested but also the resulting fabric before it could be used in a garment labelled 'Crimplene'

The dynamic impact on the industry was remarkable; it galvanized a whole chunk of the British textile industry which had already begun to look fearfully towards the Far East as a future threat. They were right to be concerned but 'Crimplene' bought some time; and for a few years it seemed that the good times would merrily roll-on for ever. But success, as so often carried the portent of decline. By the late sixties 'Crimplene' reached nearly 50% of the double-jersey market which itself had grown so dramatically in the women's clothing market. A director of M&S wrote in the Textile Institute Polyester 50th anniversary review 'There was no argument that the ICI promotion of Crimplene caught the imaginationn of the shopping public. There has never before or since been anything like such an overnight success'. The beacon of 'Crimplene' had also inspired producers of other fibres, such as 'Courtelle acrylic' to target the sector; and there was also something of a fight-back by natural fibres with wool deploying its 'Woolmark' branding programme which it had developed as part of a general strategy against the synthetics.

But 50% market share means that the product loses exclusivity and eventually its fashion appeal; the initial one-upmanship of being able to say to friends 'I have one of those wonderful new

'Crimplene' dresses' was lost and somehow or other quality becomes undermined. Not in the genuine branded 'Crimplene' merchandise but in unbranded polyester knock-offs. Gradually it lost its excitement and fashionability with an almost inevitable decline and fall. This was accelerated by the attitudes of the increasingly important youth market, with a strong trend to a natural somewhat 'scruffy' look. Originally 'Crimplene' had jelled with the swinging sixties, even for the younger market, attracted by the ultra-modern geometric styling of fashion revolutionaries like Mary Quant. But fashion being fashion moved inexorably on... In due course the damning expression 'mums in polyester suits' pushed it on the fashion scrap-heap. At the same time competition in textured polyester was growing rapidly and apart from losing what had been major export business ICI had to face up to a lower share of the British home market. One positive response was to restructure the business model. A new company was formed – called Intex – which combined forces with some of the licensees, introduced new texturing technology at manufacturing level, and effectively controlled 'Crimplene' marketing and distribution. A wider range of yarns was developed for much finer fabric types; also heavier yarns for traditional type knitwear. Intex represented more than half of textured yarns produced or sold in Britain.

An Attempt to diversify into menswear never really took off despite the introduction of 'dull' yarns more suitable for the masculine look. One brave approach by ICI to introduce a new brand 'Crimplene for Men' was to link it to a new concept for men's suits – 'The Easy Suit'. Special styles were commissioned from well-known menswear designers to create unstructured, unlined relaxed suits to resonate with the modern age of get-up-and-go. They were supposed to be right for business wear as well as for leisure. However this type of garments did not lend itself to the multifarious traditional pockets of a man's suit, perhaps the male equivalent of the female handbag, This point was picked-up by the 'Financial Times' when reviewing the 'Crimplene for Men' story. They had a little fun with the suggestion that men could have handbags too. In fact at that time some trend-setting business men were using a masculine version of the handbag, mainly inspired by their continental cousins, principally French and Italian who were in no way fazed by negative comments about being overtly feminine. However it does seem to have faded away even in Rome and Paris being replaced by the attache case or shoulder satchel. Handbag?

For a time it looked as if a warp-knit 'Crimplene' might catch on. This employed a creative way of giving the fabric a woven menswear look by heat transfer printing using classic pin-stripes and checks. However it started in the lower end of the market, got stuck there and eventually petered out.

Trousers on their own – not as the bottom half of a suit – were more successful and in the late sixties were selling at around three million pairs per annum. A lot of this business was through chain stores which were showing signs of the dominance which would be achieved later and continues today. In the mid fifties they controlled around 30% of all clothing sales which by mid seventies was up to 50%, and more.

In rounding up the polyester story it is worth mentioning the internationalization undertaken successfully by ICI. A licensing programme first brought in continental manufacturers then Japanese licensees. There was also an arrangement with the communist Soviet Union to transfer technology including building a plant in co-operation with UK engineering groups with start-up support. It established production through its subsidiaries in Australia, Canada, India and South Africa. A plant was built in Portugal in a joint venture with a Portuguese textile company. This was a shrewd strategic move because the strong Portuguese textile industry was thought to offer the best means in Europe of competing with growing low cost imports. It did at first with fabric exports to Europe, later in garments.

Perhaps the biggest overseas move was the joint venture with American Celanese to build a manufacturing facility under the company name Fiber Industries, taking on Du Pont on its home ground.

Some of the ICI licensees in Europe, for example Hoechst, also began a programme of expansion overseas.

Polyester remains one of the world's most important textile materials, with a major share of synthetic fibre production. No longer thought of as a branded product but more a basic material – although there are niches such as specialized sportswear in which branding still plays a part.

Polyester, other than in textiles, has wide usage today including the ubiquitous PET, polyester bottle, standard in the soft drinks industry.

And development is still taking place.

4

The acrylics and others make their play

OK tick the box for nylon, tick the box for polyester....the other major contender in the synthetic fibre contest to rule the textile world was the acrylic family – a family because there were competing processes to manufacture an acrylic fibre. The goal was the same: a fibre which could replicate the characteristics of wool with bulk, warmth. soft texture, extensibility and 'character'. The nature of this target steered it into knitting rather than weaving; and to be more specific it was destined for the big growing knitwear market: sweaters, jumpers, cardigans. In time it also found its way into circular-knit jersey fabrics, but there it came up against some stiff competition from polyester. In knitwear garments the only real competition was good old fashioned wool. Polyester and nylon never really made it in the fully-fashioned knitwear garment market because their characteristics did not fit although some valiant efforts were made to overcome this. Fibre producers seldom gave up trying to fit their particular fibres into as many uses as possible. Du Pont's strength was that having nylon, polyester and acrylic in its portfolio, it could place them in the most suitable business areas to optimize their individual properties.

Although Acrylic Was the last of the 'big four' on the textile scene-after rayon, nylon and polyester – the technology was not brand new. Investigation into acrylic polymers had been proceeding almost in parallel with nylon and polyester throughout the thirties and forties, although at first focused on plastics rather than textiles.

It was not until the fifties that it began to feature in the 'consumer apparel market' to use American marketing jargon of the time. And it was in the US under Du Pont's 'Orlon' label that sweaters first appeared on the knitwear counters of leading US stores, starting in the high end with a classic launch programme supported by magazine and newspaper advertising, sometimes jointly with stores and knitwear manufacturers. The ambition was to portray it as a credible alternative to cashmere which at that time was the fine knitwear 'par excellence' but priced outside most customers' pockets. After up-market positioning had established a luxury image for 'Orlon' it was then offered and promoted as a product to compete with lambswool in the medium volume market.

It did not do a bad job. It had a warm soft handle and dyed well to give some very attractive colours: pale and subtle or strong and bright...a great scarlet! The added benefit was that these colours had a high fastness to washing. Furthermore because of its thermoplastic nature garments retained their shape after frequent washing, a superiority over wool which was valued by consumers both in fashion terms and for practicality. The sweater girl was in vogue, and she stayed popular for quite a while in the shape of film stars like Jane Russell and Marilyn. But who was the original 'it' girl? Could it have been the Hollywood movie star Ann Sheridan?

In The US 'Orlon' had to face up to new entrants such as 'Acrilan' and 'Creslan'. As acrylics moved increasingly into the mass market the quality levels came down and the principal defect became more apparent: a propensity to 'pilling'. This was the term to describe the build-up of little balls of fibre fluff which clung to the garment in an unsightly way. This problem was to prove to be a bane to synthetics. The problem, was that synthetics are much stronger than wool. Wool also produces a 'pilling' characteristic but whereas a single synthetic strand keeps the fluff ball attached to the garment the weaker wool fibre snaps and the 'pill' drops off. Whatever the fibre, in wear and washing the friction rubs up the fibres and causes 'pilling'.

Various remedies were developed, special finishes and modified fibre and fabric structures; but 'pilling' was a real problem... and in truth still is but not only in synthetics. As natural fibres were modified to ape synthetics' easy-care properties they developed some of the synthetic bad habits…

Another negative was static electricity. Rubbing against the fabric in wear stimulates electric charges which cause the garment to crackle and stick to the body. Again wool was not immune to this problem but customers were more forgiving of old fashioned wool, but expected more of a new 'wonder fibre'. Special finishes and changes to fibre cross-section brought considerable improvement but to some extent the damage was already done in the mind of the potential customer.

In the UK the major indigenous player was Courtaulds with its 'Courtelle'. Although as with the other acrylics it was aimed at knitwear it also featured in the double-knit jersey sector for dresses, trouser suits and jackets. This type of garment is known as cut and sewn.

Another acrylic made in the UK was 'Acrilan' . A brand of the American group Chemstrand. Acrylics are also used in hand-knitting yarns ('wools') and upholstery.

As Part Of Industrial change textiles were migrating far, far eastwards and now acrylic production is centred in the Far East.

New garment types such as the Pashamina have suddenly appeared on railway station outlet stores as if from nowhere…an example of sudden unplanned popular fashion. And it is used in multi-fibre blends.

There is acrylic production in America but for special use as a precursor for the manufacture of advanced technology products such as carbon fibre used in

Formula One racing cars.

Another Synthetic which has revolutionized some types of fabric and clothing is spandex, otherwise known as elastalane or elastomerics-all generic terms. However the generic term spandex is interesting and a bit of a giveaway to cross-word puzzlers: it's an anagram of 'expands'. This synthetic is in effect a substitute for rubber, although substitute is perhaps an over-derogatory term because spandex performs better than the natural product. It has exceptional elasticity and it is stronger and more durable than rubber.

Spandex was discovered by Joseph Shivers at Du Pont and branded as 'Lycra'.

Unsurprisingly with its unique properties it is used in many types of active sports' clothes. Swimwear, cycling shorts, athletic tops and shorts and gymnasts outfits. The sleek tight body fit bestowed by a high percentage of spandex gives the advantage of streamlining, vitally important in today's sporting arena where every tenth of a second counts. Athletes value it because although smoothing body contours it also permits total freedom of movement. A classic one hundred metre Olympic sprinter clearly demonstrates both of these features, male and female.

It has had an interesting career too in the world of entertainment providing a sleek fit for TV superheroes. It was also used in the seventies to provide stars of 'rock' and 'heavy metal' bands with similar benefits when prancing and leaping about on the stage. For them it's slinky shine was also a bonus. Country and western singers also adopted it and one may ponder on the extent to which Miss Dolly Parton tested its outstanding modulus of elasticity.

Active sports clothing uses 100% spandex or high level blends; however for everyday clothes small amounts can give a modest degree of stretch for comfort. You find it in blends with wool, cotton or rayon to give the appearance of traditional natural fibres.

After The Arrival of the three major synthetics polymer research did not freeze but it did become more focused, more specialist. No one seemed to have hopes of another block-busting winner. There were some notable niche developments. Apart from spandex there were other valuable fibres, two more in fact from Du Pont. These were from the polyamide related aramid chemical group.

Nomex gave us the first real form of protection against flames and intense heat. It is used in fire-fighting suits and in protective clothing for motor racing drivers.

It has relatively low strength but as well as its unique heat resist property it has good electrical insulation, and chemical and radio-active resistance. A very important all-rounder for the modern world.

The other Du Pont fibre also based on aramid chemistry is 'Kevlar' It is used to make fabrics of exceptional strength, five times stronger than steel on a weight for weight comparison. Although it features in many fields where super-performance is needed, it is possibly most famous for its use in body-armour, part of that contemporary industry providing protection which our developed society seems to require. We see it all the time on television worn by the military, the police and

56

war-front journalists symbolic of the dangerous planet on which we live. Shall we reach a stage where it becomes standard civilian wear for anti-terrorist protection?

Carbon Fibre is another high-tech invention which is used as a reinforcement in high performance polymers mainly for rigid structures. It is very strong, light and expensive!

Various forms are used in aerospace, power-boats, racing cycles, tennis raquets and golf clubs. One of its most spectacular appearances is on Formula One racing cars. Many people will have seen shards of carbon fibre splattered over the track in a 'racing incident'. F1 technology exploits the fact that carbon fibre can be immensely strong in one direction, for a load bearing structure, and is yet relatively weak in another dimension thus shearing in some accidents.

The Last Synthetic to include in this round-up is polypropolene, an olefin derivative. It is a strong fibre with good elasticity and shock strength. Despite attempts to position it as a material for knitwear it never succeeded and it found a place in outlets where its physical characteristics were most appreciated: carpets and upholstery for its abrasion resistance; and especially in ropes and twines for its resilience. Merkalon was one brand name but it never achieved serious recognition.

There are a number of other fibres, new and modified versions of existing ones, but these should give a good idea of the main ones and the way they evolved.

(left) *Presentation of new knitwear designs and trends. Typical or seasonal merchandise promotions.*

(below) *An earnest discussion about the future of the textile industry... or something!*

(above) *ICI Fibres 'Festival of Fashion' Dublin, 1966.*

(right) *Fashion focus event ICI Fibres Knightsbridge.*

(above) *Newspaper
fashion report featuring
'Yorkshire Rosette'
Terylene tweed in latest
style long topcoat with
matching short skirt.*

(left) *Appointment of BNS
advertising agent in Dublin,
Peter Owens (centre); left Joe
Scott BNS and Courtaulds'
universally loved agent in
Ireland.*

(above) *Fashion show ICI Fibres Dublin office.*

(below) *World famous swing violinist Stefan Grapelli features in printed knitted Terylene shirt promotion at Dorchester hotel.*

(right) *Planning BNS strategy in Germany.*

(left) *'Hands on' quality control!*

(left) *Putting on the 'style' – Kipper ties have it.*

(right) *Tribute to Mike Page – BNS/ICI – on his retirement. Sadly he died October 2010.*

SOLD EXPORTS AND SAW THE WORLD

British Nylon in India

(left) *Report by Mike Page on BNS participation in Indian Industrial exhibition, Delhi.*

ICI's nylon men reach the fork in the road

(right) *Sunday Times business news report about ICI Fibres deputy chairman Howard Morris (end of era).*

(below) *ICI Fibres Dublin office team 1968.*

(left) *ICI Fibres Post announcement of management restructuring.*

1967 STARTS WITH REGROUP

APPOINTMENT

(above) *A cast of characters – ICI Fibres.*

5

Textile barons and retail kings

The textile trade in Britain has always involved powerful men often with large egos. It started back in the 18th century when the owners of water mills in small communities were able to develop the early forms of power assisted weaving The mood of the age motivated the adventurous to invest, to take risks. As the industrial revolution rapidly gained momentum in England the entrepreneurs of the day began to speculate and put their cash into technical innovation which was revolutionising textiles in the spinning and weaving industries. Of course this was to some extent risky because the technology was new and untried. The Mule, Spinning Jenny and the Fly-frame Shuttle were a big step forward from hand spinning and weaving but new, so it took courage to back them. There was a growing market for fabric both at home and overseas with captive markets in countries of the Empire. But the new machinery was expensive and only the successful ones were able to afford the cost of investing in it and in the building of new factories. Bigger investment resulted in more economic production and greater market dominance. The rich got richer… as ever.

In centres for the increasingly important wool business such as Bradford and big Lancashire cotton industry in Manchester, a new aristocracy was born and dynasties begun. There were sometimes occasional glitches in succession. The founders had been big bold men, hard working and despite their business risk-taking, prudent in their personal finances. Some of their heirs in subsequent generations spent too much on personal luxury and not enough in developing the business. Too much money spent on Bentley sports cars and not enough on new plant.

Nevertheless many of these companies continued to grow. During the years of the Great War, 1914-1918, a lot of money was made in manufacturing uniforms and other wartime materials. They also survived the Great Depression and went through the thirties into World - War Two.

After the second world-war a number were able to profit from the post war boom with a world demand for textiles and clothing although the UK's economic straits meant that much of the production went to export. And so quite a few of the textile groups with pedigrees going back to the 18th century were set to play a major role in the synthetic textile revolution which began in the late forties.

Salts of Saltaire were major 'top makers' blending wool and 'Terylene' polyester into a form which could be used to spin yarns for weaving or knitting. Sir Titus Salt had created Saltaire model village for his workers on the outskirts of Bradford.

Another famous firm Listers, at one time known as Lister Aked, began weaving silk velvet in the 1800s. At one time it was the largest silk weaving mill in the world. Its chimney could be seen from virtually anywhere in the Bradford area. It is now preserved as a grade two listed building celebrating its unusual, for an industrial building, Italianate architecture. During World War Two Listers made 1330 miles of parachute fabric. Aked under chairman Eugen Kornberg was one of the pioneers of Terylene/worsted fabric linked to the permanently pleated skirt.

A Knitting Company Which Was to become very important in the synthetic knitwear was Corah. Nathaniel Corah had begun in the early 1800s using a hand-knitting frame which he had set up in a farm shed, after training as a knitting-frame smith. Apart from being a good craftsman he was clearly a shrewd entrepreneur. By 1866 he had more than a thousand workers at his factory near Leicester. Corahs enjoyed a long term relationship as major suppliers to Marks & Spencer.

Essential players in the early stages of development of continuous filament synthetics were the throwsters or yarn processors. By applying twist to the early forms of nylon yarn they made it acceptable for commercial production. Later when this stage was no longer necessary because fibre producers had introduced new technology obviating that step, they were a key factor in the commercialization of crimped yarns giving bulk and stretch to take synthetics into new end uses.

Major companies in that field were GH Heath and William Tatton who had a long tradition in silk processing later featuring as licensees for 'Crimplene'.

In The Double Decade 1960-1980 one of the leading figures was Joe Hyman. He played a key role in transforming the British textile industry. If some of his more ambitious merger plans had succeeded he might have created an even stronger super-group which could have sustained a major textile manufacturing industry in Britain longer than it in fact survived.

His career began with a somewhat unlikely move. Wm. Hollins was a famous English textile company with roots going back to 1784 in the town of Pleasley on the Derbyshire/Nottinghamshire borders. Hollins was known for its Viyella fabric – a blend of wool and cotton which had a distinctive soft warm handle or feel. It was very popular for nightwear and gentlemanly Tattersall check shirts, perfect for country gents and weekend sportsmen. But a bit pricey...Clydella was a less expensive version.

The directors of Wm. Hollins realized that they needed to re-energise the business, looked around, and came up with a dynamic young Manchester fabric merchant Joe Hyman, who had made a bold move in setting up a progressive warp-knitting plant – Gainsborough Cornard – near Sudbury. The board invited him into the business. Good choice – except that within a very short time he was running the business and most of them had retired. Perhaps it suited them that way.

This special type of reverse takeover was the beginning of a programme of revolutionary restructuring of the UK textile industry. Hyman moved swiftly to create an empire of textile companies with more than 40 factories around Britain covering a wide spread of manufacturing activity. Early on Joe Hyman saw the opportunity offered by synthetic fibres and they featured strongly in his development plans. Warp knitting nylon to make shirting fabrics and bed sheets; polyester and nylon filament weaving; polyester/rayon blend weaving for trousers; and dyeing and finishing plants to process these fabrics.

A key element in Hyman's strategy was vertical structuring: he wanted to control manufacture from the basic fibre level right through to the retail sale point, managing the complicated textile trade pipeline. So he had his own spinning facilities at the start of the pipeline and manufacture of garments at the other end involving leading shirt brands such as Van Heusen, Peter England and Roccola.

It is interesting that the Viyella remained the central name throughout the main stages of Joe Hyman's empire building: Carrington Viyella, Vantona Viyella, Coats Viyella...This was appropriate because the original Viyella was the first branded fabric in the world. And possibly became his talisman.

Reports and recollections suggest that Joe was something of a tyrant – or shall we say a dominant personality. No alcohol allowed in the directors' dining room at his Savile Row headquarters; Schloer apple juice was the preferred tiple although one might sense that some of the old guard from traditional textile backgrounds might have relished a glass of something stronger to get them through a tough session with the 'boss'. Exercising the level of hands-on and strategic management which his creative style required was very demanding: reputedly he had three personal secretaries each with individual skills.

He was able to pursue the dream of building a strong British textile business largely because of the support from British chemical giant ICI. It was clearly in their interest to have a strong home – almost captive – market for their fibres.

ICI's earlier attempt to take over Courtaulds – which as well as man-made fibre production had its own spinning, weaving, knitting and garment making – had resulted in the creation of the ICI Fibres, bringing together nylon and polyester. Working with Joe Hyman and the Viyella group was another way forward. It is interesting to speculate whether if ICI had succeeded in its bid for Courtaulds, the second stage might have been to put together with Viyella the fabric producing and garment making activities of Viyella and Courtaulds to create perhaps the biggest vertical textile group in the world. There could have been some interesting personality clashes between people like Joe Hyman and Frank Kearton the strong man at Courtaulds who fought the ICI bid. But it would certainly have been stimulating stuff!

Another significant name was Sir John Harvey Jones who at the time was the main board ICI director responsible for ICI Fibres division and for liaison with the Viyella group – essentially with Jo Hyman. Harvey -Jones was well able to deal with Joe and with other decision makers – for example Marks & Spencer. He was very different from the then archetypal ICI director. He was not a scientist nor even a graduate. He was an ex Royal Navy submariner with a very positive individual salty style. His highly individual dress style was notorious: strange green suits, kipper ties and high heeled shoes. He went on to become one of ICI's most successful chairmen and later the TV Trouble Shooter. Anyone who worked with John held him in great respect not only for his outstanding business talent but also for his management style and his humanity.

The juggling for position amongst fibre producers and textile manufacturing groups in the UK certainly played its part in the way the man-made fibre sector developed and influenced the clothing and textile markets. Changes in the retail scene had a profound affect on consumer purchasing patterns and established a model which continues today. Merchandise available in volume at competitive prices and with interesting promotional story opportunities suited the expanding chain variety stores such as Marks & Spencer and British Home Stores. It also fitted in with mail-order catalogues for example Littlewoods.

From the fibre producers' perspective it was valuable to have an expanding mass market linking with increasing fibre production. Inevitably there was a problem with branding. Fibre producers were keen to establish their own trade-names – Bri-nylon, Terylene, Crimplene, Orlon – whereas leading chain stores were actively working to develop their own brands. In fact it worked out pretty well. When M&S

could see that a fibre brand was being extensively promoted on a national scale they accepted that it was in their own interest to take advantage of this and deploy the fibre brand on their merchandise and for in-store promotion. A good relationship developed at various levels between ICI and M&S to co-operate in identifying new consumer clothing requirements and exploring the technology to meet these opportunities.

M&S Were Central To A Retail Revolution and played a leading role in the way synthetic fibre products were marketed. Louis Goodman who was a long serving board member of Marks & Spencer had a key role in working with fibre producers and suppliers linking fibre, fabric and garment. It is said that Joe Hyman once described M&S as 'manufacturers without machinery'.

According to Louis Goodman the M&S policy was set by Lord Marks when he told a shareholders meeting in 1945:

"Science is producing new materials and new processes as well as improving existing materials. In our own field new synthetic fibres and plastics are being created which will generate additional demands and wants by the public. A distributive organization like ours, in direct contact with the large public and in close association with manufacturing industry, can help to expedite the development of such new materials" Prophetic or what?

In an article in a Textile Institute publication Louis Goodman wrote 'It was an exciting era for creating new products and watching trial lines grow into multimillion pound ranges. By 1967 about fifteen years after the first appearance of Terylene, about two-thirds of St. Michael merchandise was made from man-made fibres'.

In Menswear the tailored suit market was still very large in the 60s and 70s, and the multiple tailors epitomized by Burtons, Jacksons, Prices' Fifty Shilling Tailors, Alexandre, and Colliers had the bulk of that trade.

These outlets were selling something like nine million suits each year, with Burton's Hudson Road factory in Leeds reputedly employing 10,000 workers in it's prime. Ironically it was said that they did not make their money selling suits. They made money by taking the deposit, say £5, and having it to invest for six weeks until the suit was ready for collection. At the same time they got up to one hundred and twenty days credit from the fabric suppliers. The suits were made-to-measure from standard blocks which suited most of their customers' sizes. Customers chose their fabrics from pattern books. It was a competitive business with the Multiple tailors having stores on the high-streets of Britain pretty much cheek to jowl. Apparently one trick to pull-in customers strolling the high street on a Saturday afternoon was to use 'nudgers'. The store entrances often had, by design, plate glass windows which cunningly curved in towards the store entrance. If a likely lad paused to have a quick look in the window he was gently nudged round into the store and before he could escape found himself at the counter ordering a suit! Possibly there was lack of confidence in young working chaps of that era which prevented them from making a fuss. They found it less embarrassing to go with the flow persuading themselves that they had intended to buy a new suit anyway. Or perhaps it is an urban myth. Whatever, the multiple tailors were very important outlets for fibre producers which supported them with promotional subsidiaries and fabric developments and designs to suit their specific customers.

The power of large retail outlets was manifest and growing...but of course there were still many smaller chains-like Austin Reed – and they were also promoting the new synthetic fibre garments, mainly in blends and in a more upmarket way befitting their market status.

It is fascinating to see how names with heritage right back to the eighteen hundreds were involved in the up-to-the minute science based fibres, adapting and modifying to exploit the opportunities they represented.

It Was a complex period in which synthetic fibres played some contributory role in the major changes taking place in the British textile industry and in the nature of retailing and distribution. Synthetic fibres also changed the way people, consumers, thought about clothes: how to buy them, where to buy them, what to expect of them; and to think of clothes as fun – minor luxuries – rather than necessities.

The Advertising Game

In the years between the wars advertising had begun to move further from the simple form of providing an advisory service to tell consumers at which shops and stores products could be purchased locally – to a more sophisticated activity launching new brands and defending existing ones. Advertising was becoming a marketing tool rather than a form of information service. There were already some established trade names: Sunlight soap, Cadbury, Pears soap, Ovaltine, Players were a few of them. The famous Guinness poster advertisements were leaders. Media was limited: newspapers (black/white), posters, some magazines offering expensive colour; and early forms of 'pirate radio' – Radio Luxemburg with Big Bill Campbell.

After World War II it took some time for advertising to crank-up because potential advertising clients had only a limited amount of goods to sell and questioned the need to 'waste money' on advertising. Some more enlightened marketers did invest in brand building for the future.

By the time the man-made fibre's wagon was rolling, the advertising industry was ready for it – and welcoming this new exciting business opportunity. The prospect of a business which would place so much emphasis on colour and creativity was enticing. And the developments in media, especially of course television, fitted in well although this happened much sooner in America than in Britain.

Initially fashion magazines like Vogue, Harpers were central to synthetic fibre advertising – glossy black and white plus full colour pages – as that process was developed. Later the colour magazine supplements to Sunday Newspapers opened up an opportunity for colour in publications reaching much bigger readerships than fashion magazines. Specialized merchandising linked paid-for advertising with editorial so called 'advertorials'. A special feature was prepared by the publication which because it appeared to be editorial was reckoned to be more persuasive from the reader's perspective than an obvious advertisement. Often these were tied-in with selected top retailers with the luxury association the advertiser was seeking.

The Increased Popular Interest In Fashion was encouraging a breed of specialist journalists writing on the subject of fashion: reporting on fashion shows, new styles and fabrics. This in turn led to what was to grow into a major commercial activity-public relations, although at first it was more often known as press relations.

One of the personalities in the field was Margaret Reekie, known as Maggie. She was the press officer for British Nylon Spinners right from its beginning. It was a key job presenting not only a brand new product but a brand new company in totally new industry. She had a good relationship with the leading fashion writers such as the influential Alison Settle. To see Maggie in action at a press event was really something .She had the ability somehow to corral a bunch of journalists and put them in a corner so that she could talk to them as a group. Definitely a great character. She cut her teeth as an information officer at the Ministry of Information during the war. She worked on two campaigns which she felt were in conflict. One was the knickers and nightdresses project to encourage women to use less fabric for their underwear reducing garments to the minimum. The other was an anti venereal disease scheme!

As Competition between fibre brands intensified, very large budgets were being allocated to national advertising programmes – perhaps £3-5 million for a seasonal campaign by just one fibre producer. Bearing in mind the profit levels at the beginning, such budgets were not unreasonable.

Although there was a genuine target of persuading consumers to go into stores and ask for specific brands, or at least re-act positively to them when they were seen in displays or on retail shelves, an underlying objective was to use the power of the advertising to influence the textile trade. This was done in various ways. One was to pre-advertise in the trade magazines setting out specimens of the forthcoming mainstream consumer ads together with a strong portrayal of the advertising schedule. This was to persuade manufacturers and retailers to produce and stock garments carrying the promoted brand. Another method was to use sales and merchandising teams to make presentations of the consumer advertising schedules to their trade contacts. So salesmen selling fibre to spinners or yarns to stocking manufactures would have copies of the ads to show their customers. Further down the long-textile pipeline, merchandisers would talk to garment makers and retailers to encourage them to benefit from the advertising by using the fibre brand names on garments and promoting them in the stores.

Fibre-producers and their adverting agencies were actively working to improve knowledge of nylon and its widening range of end-uses, deploying alternatives to straight forward advertisements. An early example was the 'British Nylon Fair' in 1959 principally targeting all levels of the textile and retail trades. Subsequently a similar event was used as part of the launch of Bri-nylon, the British Nylon Spinners (BNS) landmark scheme to brand a generic fibre. At that time large promotional fairs such as the Ideal Homes Exhibition were much in vogue so the concept was familiar to the trade and public. In the early sixties the british Nylon fair became the centrepiece of a London social week with some garment makers doing their own related shows, not unlike the contemporary London Fashion Week.

Similar events ,which included fashion shows, were

rolled out by BNS in main centres across the UK, so called 'town promotions'.

One event outside the UK was in Dublin an important market for BNS. It was at the Shelbourne Hotel on College Green in the centre of Dublin and attracted considerable interest. Invitations to the fashion shows were given out by participating shops and stores, some fifty being involved in having Bri-nylon instore promotions.

Before the fashion shows there were a thirty minute question and answer sessions-the 'Bri-nylon Brains Trust'. A number of questions were placed in the audience to encourage people to put their hands up. Possibly one of the best was a question delivered in a somewhat suspect Irish accent "Could one of you fine gents ever tell me why is it that them ould Bri-nylon socks are so comfortable to wear and last so long?". The only answer of course was "I'm glad, sir, that you asked that question...the reason is...."

Unsurprisingly there were more than a few great characters in the Irish textile scene. For example Harry Keogh, owner of a nylon warp knitting factory and the race horse Royal Tan. He went on a trip to the Bri-nylon Fair organised by the much loved BNS agent Joe Scott. Almost as soon as the Aer Lingus plane had left the Dublin airport tarmac Harry pressed the button for the stewardess. "Bejasus I could murder a G&T" he said. Joe Scott sitting next to him commented "I thought you'd given up drink for Lent", "Oh I have" replied Harry "In Ireland, but now we're just about off the ould Irish sod".

There was also the managing director of a large Irish processing mill whose hearing aid only became operative when, during a price negotiation, the cost had fallen to its lowest point.

To convince the chairman of a leading Irish textile group to become a registered Bri-nylon user, an inovative low cost programme was devised. A dozen full size billboard posters were carefully sited on the route he drove daily, in his Alvis sport coupe from home to work and back. Full page ads were placed in the daily newspaper which was always on his desk every morning, every other day for two weeks, and regional TV ads, at modest cost, were shown. All of these media had the same simple message "Buy Irish Bri-nylon" shown on a blow-up of a giant swing ticket. Within a short time he had signed up.... 'this must be one of the biggest advertising campaigns ever, wherever you look you see the name Bri-nylon'. In fact it was a positive response on his part and created good business for his group.

Two Examples of the innovative and complex promotions in the seventies were very different and yet had key similarities.

One programme was for Evvaprest fabric intended for trousers, operated by A&S Orr, part of the Viyella group. It had the bold objective of creating a fabric brand which could be transferred to a garment – trousers – and promoted at retail level to menswear customers. It was perhaps a heritage from Viyella.

The branding was innovative but it was the underlying, quite complex, technical breakthrough which underpinned its success. Brian Hamilton who was chairman at the time still talks with pride about the concept and the technical achievement.

"The aim was to replicate a worsted trousering in handle and appearance, but also to give it the easy-care properties increasingly expected in the market-place. It was therefore a purpose designed and constructed fabric. After a number of development trials a blend of 'Terylene' and rayon was selected; both had the advantage of being available in spun dyed fibre and in fact the Terylene was specially developed for us in a range of melt-dye colours. this was important because we wanted to be able to offer a good selection of modern designs for the growing fashion market in menswear even in volume outlets. We knew that the finishing of the fabric would be critical so we took the radical step of having the cloth finished in a works which was a traditional worsted fabric processor in line with our target of a wool worsted character cloth. Both the technical and the marketing programmes worked out well and we have to regard it as one of our success stories with a lot of cloth being sold."

To a large extent it was a good example of the attractions of vertical flows with the company controlling development, spinning, weaving, finishing and promotion.

Tricopress was another unorthodox scheme to revitalize the Bri-nylon shirt business. Sales had reached a very high level of UK market penetration, at one point almost 40% of shirt sales. This was fine in the short term but the fickle finger of fashion swings all too swiftly... Moreover as volume grew prices came down and prestige shirt manufacturers like Van Heusen were muttering about the difficulty of competing with the chain stores who were selling Bri-nylon shirts like hot cakes.

The idea was conceived of developing a luxury form of fabric which would give top-end shirt firms an exclusive, but it had to be possible to distinguish it from the standard fabric used for Bri-nylon shirts. Starting from the premise that a finer, silkier fabric would look and feel different, and be attractive to the target shirt makers, a programme to use finer gauge warp knitting machines was set up; the clever bit was that there were very few such machines available and supply would be restricted not artificially but naturally.

The next step was a unique promotional strategy. The small club of knitters involved agreed to include a small surcharge in their fabric price which would be rebated to a special fund for specific 'Tricopress' advertising. The shirt maker would also build into his shirt price a small premium to be used in the joint promotion for 'Tricopress' And ICI would contribute to the campaign. It was ingenious and successful despite somewhat complex administration to work out who owed what to whom! Still it bought time in a gradually decaying nylon shirt market.

The Synthetic Fibre industry and the advertising business were continuing to seek new paths through the burgeoning consumerism which was changing many established ways of buying; and in particular trying to understand what customers really wanted before they knew it themselves. It has been described as inventing consumer 'wants' rather than 'needs' but in truth it is a bit more subtle than that. People en masse do have latent hopes and aspirations and it was good marketing to anticipate and understand these ahead of the pack.

Marketing as a concept and as a name was just beginning to feature.There was considerable misunderstanding about its role. At one simplistic level it was regarded almost as another term for advertising or promotion; in fact it was much more significant and represented a major shift in the approach to business. It introduced a scientific approach to selling with research and forecasting techniques instead of over reliance on talent and hunch. Ed Bursk professor of marketing at Harvard Business School described marketing pithily as 'Running your business with the customer in mind' and it still takes a lot of beating in expressing the essence of it. It was Harvard Business School in 1926 which introduced the world's first formal programme to teach business as an academic subject. This began a process which now virtually demands that any contender for a senior business post must have an MBA (Master of Business Administration). It was an important change from the traditional English attitude which saw a good degree in history as a sound educational platform for commerce: business was not seen as a profession in the same way as law. You more or less learnt it on the job.

One of the new tools of marketing was research and this was especially interesting to those engaged in marketing a new type of product-synthetic fibres. Again there was synergy with the advertising business. As their budgets became larger, almost exponentially, so it was essential for them to be able to demonstrate to clients that their money was being spent wisely and effectively. Methods used became more sophisticated: pre-testing different forms of advertisement using small consumer panels, couponing advertisements to evaluate the response to alternative graphics and copy; and measuring response using specialists to interview randomly selected respondents. The advent of TV brought new techniques such as TAM-television audience measurement and the intriguing 'flush factor' measuring so-called comfort breaks while the ads were on.

Facts and figures were one thing, but valuable though they were both the agencies and their clients wanted to know more, to get into the mind and understand consumer attitudes, how they were influenced by this or that form of advertising. It was something of a field day for trained psychologists who took their places alongside the statisticians and market economists. The new big thing was attitude research and the leading guru of the time was Dr Alfred Dichter with his research centre at Croton-on-Hudson, New York State.

He did some work for Du Pont on nylon stockings in the sixties and came up with the report that women regarded a ladder in their precious nylons as being akin to rape. It is by no means certain but the story has it that the research may have led to more technical research resulted in 'ladderstop' stitch constructions. He also gave the opinion that consumers preferred synthetic/natural fibre blends – the character of the natural fibre like cotton and the performance of the synthetic. Not really an illogical conclusion and hardly surprising but not the one which fibre producers wanted to hear: their dream was still the 100% synthetic garment which they persuaded themselves was what the customer really wanted, once their fibres had been further developed.

The Book 'The Hidden Persuaders' was a popular title in the 60s with some growing – and seriously over rated – concerns about the power of advertising and how it could perhaps penetrate minds and motivate undesirable actions. The idea of subliminal advertising was also a modest issue: the proposition that a deep message could be layed-in behind a normal film or TV programme so that while watching and enjoying you would be soaking up hidden messages. A sinister thought in terms of political campaigning which was anyway learning how to exploit the power of television.

It wasn't only ad agencies who undertook attitude surveys. British Nylon Spinners carried out a study into the use of sex in advertising. How would the average consumer react to explicit sexual implications? A number of photographs showing young women in various items of underwear were shown to a statistically designed panel. They were asked to rate the photos in order of preference. At one end of the scale was a winsome, modest but very pretty girl next door holding a rose bud to her lips .At the other end of the scale was a dark eyed but blonde haired temptress with her legs wrapped round the post of a four poster bed and her nightdress in some disarray. It was a surprise then to find that women rated the chocolate box sweetie picture as number one; while most blokes seemed to prefer the strumpet in tight knickers. Today though she would seem to be rather demure.

It was one of the great periods in the advertising world. 'Madison Avenue USA' was one of the best books in the 60s to tell the tale of button down shirts and the big players in the ad world. A story being graphically re-told in the excellent American TV series Madmen. It was also a tremendous time in the synthetic fibre industry, making huge profits but still learning its trade. At the time no-one realized that the golden days would sometime have to fade.

A form of below the line advertising as it used to be called, became increasingly important on the International scene. As the battle between fibre producers intensified one of the main promotional efforts focused on the area of major textile fairs which were attended in large numbers by garment makers, retailers and infuential journalists writing for textile and fashion publications around the world.

In Europe two of the leading exhibitions were in Germany. These were Interstoff; a huge fair to promote fabrics (and indirectly fibres), which was held twice a year in the spacious exhibition halls in Frnakfurt. Anyone who was anyone in textiles was there. 'HAKA' was the menswear fair at Cologne where the latest styles in the new seasons fabrics were presented. Again a good show -ground for the fibre makers to display their latest developments. At both events - and others - the fabric people vied with each other to make the strongest and most persuasive statements. Stand designs became more esoteric and hospitality more lavish... Fashion shows with top models strutting their stuff on the catwalks were increasingly daring and spectacular. Lingerie shows were by invitation only and significantly over sub-scribed.

Paris too had an influential 'ready to wear' fair, notorious on the down-side for the impossibly long taxi-rank at the exit with with much 'Non British' queue jumping in evidence.

On Harrogate there was a major mens-wear fair, which attracted foreign buyers. Possibly attracted by the strong UK image for mens clothes and shoes. ICI Fibres considered this their back-yard with their HQ just ten minutes away.

There were similar events in the USA - for example in Dallas, but geography made them more parochial – if that term can be applied to a 180 million market!

The fickle finger of fashion

The story of man-made fibres is intimately linked with the fashion surge of the sixties, at first for good…then later… bad news.

A change in attitudes to clothes in that era brought an open minded approach to new fabrics; it helped to take away the stodginess and an acceptance of bright fresh colours and lighter weight cloth. Men's suitings dropped in weight from around 19oz per yard to 16oz for which polyester/wool blends worked well; and as central heating became more widespread they became even lighter. What were once thought of as tropical weights were becoming quite popular especially in luxury suits. Men spent most of their time in cars, offices, hotels, pubs – even homes – and with the trend to heaters these were all pretty warm places even in deep midwinter.

Women's clothes had always paid some homage to fashion, mostly to Paris, with haute couture styles to some degree permeating down to the mass market. Dior was early in exploiting its name to produce a ready-to-wear – pret a porter – range although at the top end of the retail market. Then in the sixties London became a different kind of fashion capital with young brash designers like Mary Quant creating their own innovative styles, including of course the splendid mini-skirt. Their 'street-inspired' fashion became a major international force. London was swinging and cool – or possibly hot.

And at first this vibrant, new, young exciting fashion scene played well with the synthetic fibres which were also breaking down barriers in the fashion world.

For men it was a step change in attitudes to fashion and style. Of course there had always been natty dressers. In the thirties the Duke of Windsor shone at the 'nobby' end of the market epitomized by his eponymous tie-knot; but also in the so-called working classes some of those who could afford it liked to put on the style as part of social escalation. The Hollywood image of peddled dreams was highly influential. Cary Grant the star from Bristol really looked the part. There is a story that back in the days when a cable was the most modern form of communication (fax and e-mails were not even a glint in the eye) Cary received one from a newspaper doing a story about him. Because cables were expensive efforts were always made to reduce the number of words. So this one read 'How old Cary

Grant?' He replied tongue neatly tucked in cheek 'Old Cary Grant fine stop how you?'.

The Post World War One effect had begun to force social change, in the late twenties and thirties, surely partly stimulated by cinema which opened up new horizons and built dreams for millions. Communication reaching wide audiences geographically and demographically has that effect; suddenly the masses can see what they have been missing. The sixties represented an even more significant step change; something deeper, more structural, more symbolic of a shift in social perception. Race sex, homosexuality, religious orientation all became issues. Whatever their station in life people began to look much more classless. Yes, with a man in the best hand-tailored suit the differences could be picked out – if you knew what to look for – although it wasn't long before hand-pricked lapels could be done by machine!

At the same time there was a bravery manifest in styling. Suddenly wearing pink shirts was not a cissy thing. The coloured shirt, in plain colour rainbow hues was even to be seen in boardrooms and in civil servants' Whitehall offices. Starched white collars and classic stripes were rubbing shoulders with deep blue, lemon and pea green.

For men's suit's the clean, rather sharp look was favoured. Mohair – and later mohair type – fabrics with a distinctive sheen caught the eyes of in - vogue gents about town; and the distinctive lustre quickly found its way down the ranks into the multiple tailors. Really smart for a young sport to impress his bird.

Suit styling was also being transformed. The lapels on

jackets became wider and ridiculously wider, to the point where they could move in only one direction – to start narrowing into the disappearing point.

In trousers there was a dramatic widening reaching ultimately bell bottom plus proportions, recalling the Oxford flannel bags of the twenties. As well as jolly jack tars. Widths went from 'teddy boy' 12 inches to a generous 30 inches evoking memories of Michael Caine in 'the Italian Job' being pursued along the King's Road wearing 'the gear'.

Back in the Sixties and Seventies the tie was an important item of mens clothing. The hat a standard item in the Forties and Fifties was on the way out. Amazing whwn one sees old newsreel - black and white with an almost 100% headgear coverage as blokes came up out of the ground on the way to work. But ties kept their status although with style changes in keeping with jacket lapels and trousers were slim fitting; then kipper wide when lapels had bevome ultra - wide. Ridiculous but we wore them. Polyester was a good material for ties: crease resistant and good drape. It could also be produced in a range of patterns and colours.

To some extent the tie survived because it represented a way for a man to show individuality. Even today with open neck shirts in Vogue, it can be a choice for a special occasions.

Just as a side issue it is worth remembering how men's hairstyles also fitted the new mood. Even ex-Guards officers could be seen with sideburns (sideboards?) well below the ear-line, somewhat reminiscent of Spanish bull-fighters. Long hair was anti - establishment

This was the era which popped the 'pill' bringing a

new dimension to sexual behaviour with a new permissiveness which also allowed more provocative clothing...suiting the synthetic fibres: light, rather see-through, floaty.

Not so sexy but popular on a Saturday morning in the suburban supermarket was the dreaded shell suit. This was a form of ski suit made from light weight nylon and often seen in fetching colour combinations such as silver and hibiscus pink. We must hasten to add that this was essentially a female trend. Men were not sufficiently enlightened to go down that track...

The other 'big thing' was the growth of package holidays at prices which a lot of people could afford. The new foreign holiday travellers wanted suitable clothes and synthetics filled the bill: topically attractive, non-crush packing and easy-care.

What Did Not Work very well for synthetics were the attempts to introduce major innovation in some garment types. On paper the idea of a man's suit designed to use the best characteristics of a modern synthetic, seemed right. The 'Crimplene for men Easy-Suit' with some acknowledgement to Star-Trek uniforms, was intended to liberate men from the shackles of structured formal suits: another brave free world. But it didn't work; tradition seemed to have it over comfort and modernity. And men do seem to like a multifarious collection of pockets for all their manly things. Hand bags for men didn't really catch on; that is not until the advent of the Samsonite, and similar style, brief-case which let's face it is in reality a masculine handbag.

So It Was All going Rather Well for the synthetics: social change, fashion change with synthetic fibres collaborating to general benefit.

However…along came the denim jean. Which maverick was inspired by the Wild West and denim pants with copper button-up flies. Home, home on the range with bacon and beans over an open prairie fire. Keeow! We'll never know who the miscreant was but boy, did he change the fashion world. Before long the jean was beginning to dominate the leisure market; and not much later denim jackets became popular with the young for any occasion. The jean and denim fabric have gone on to be standard wear across the world in almost every situation. A lot to answer for! In reality no one person or designer was responsible; it was probably another example of 'street fashion' exerting its natural authority deriving from working clothes along with khaki twills. Dynamic fashion.

Moreover it was egalitarian, to be seen on film-stars, rock-stars, sportspeople, and politicians at week-end to impress constituents with the new cult of 'dressing down'.

When the popular predilection for celebrities arrived, even though no one was too sure what they were celebrated for, these modern-age stars were soon to be seen wearing the denim jean, perhaps more elaborate and sequinned but still jeans. Fashion models wore jeans on the way to their work where they were to be photographed in haute couture frocks. Their jeans were frequently 'distressed' or torn-leg little numbers which cost the earth.

By the way the term jean describes a type of garment with a low yoke at the back with high set slant hip pockets. Denim is the name for a fabric which is woven with an indigo blue weft yarn over a white warp. Although many newer interpretations seem to have blurred the original basic work-wear derived garment the style continues to flourish.

Whatever... it was a hard blow to synthetic fibres at that time.

What is more, possibly influenced by the jean trend, there was a decisive fashion swing in other garments to a softer, baggier, less formal-even creased look. Over-long trousers became popular bunching round their bottoms or catching under the heel. The reach-me-down, rather than looking like an older brother's cast-off was suddenly 'it'.

It was of course much later that the rather disgusting droopy drawers' look emerged for men with the seat hanging several inches below the bottom. This has sometimes been attributed to the famous fashion designer Alexander McQueen who died in 2009.

These features went completely against the key advantages which synthetics offered, the reasons why they had been popular.

The younger end of the market was quietly rejecting the 'smart' look with trousers displaying sharp pleats and set creases; shirts with structured semi-stiff collars. Just as synthetic fibres had originally contributed to a more relaxed, less formal style-but still tailored – the latest fashion went much further – off into the distance. Even before the ubiquitous, unstoppable, jean, khaki chinos – initially derived from American work-wear – had become popular with young guys for Saturday morning wear with an Oxford shirt. Names like Ben Sherman pioneered in Britain the soft button-down roll collar with a signature higher button which allowed the shirt to be worn open-neck without the collar gaping open. Borrowed from Brookes Brothers of Madison Avenue, New York. Such a contrast to the semi-stiff attached collar.

In the case of shirts this hit Bri-nylon quite hard. One of the beauties of a Bri-nylon shirt was that with proper interlining it required no ironing and the collar automatically looked semi-stiff; but it was very difficult, more or less impossible, to get the draped soft collar look that was wanted. In another branch of the synthetic fibre business things were not so bad. Polyester with its facility for blending with cotton was able to make fabrics and garments which fitted the new casual look, although even then the advantage of the blend over cotton was less palpable.

As Ever The Trend Spread and as with women's Carnaby Street fashion it moved upwards. A reverse of the traditional direction of fashion influence and change. Of course it certainly did not destroy all conventional styling but it had a serious affect. The mighty multiple tailors – the Burtons Colliers and Hepworths – did not survive in the former model. Burtons re-invented itself, both in its merchandise and store design. Hepworths re-morphed as 'Next' slotting into an entirely different market from its previous existence.

It Would Be Misleading to claim that this draconian fashion change was the cause of branded synthetic fibres' nemesis. But it did play an important part. When the industry was assailed by other major destructive influences, it would have been better able to fight back if it had not also become unfashionable.

8

What went wrong?

The world was changing that's what went wrong. Macro economics, social shifts, politics, fashion and consumer attitudes; and perhaps partly influenced by the policies of the fibre manufactures.

From the late sixties onwards the received wisdom on the part of the fibre makers seemed to be big is best. This was especially true of polyester staple fibre seen as a high volume product mainly for blending with cotton. There was talk of 'the big engine' a symbolic term for a huge plant simply pumping out fibre in standard form. If you owned the biggest most, efficient, plant you could dominate the market by sheer weight of product...and by virtue of that the most competitive price. You could put the squeeze on competitors. Possibly that could have worked in a confined market.

The trouble was that as patents expired or became narrower, more producers came in world-wide and the market was not confined – it was global. Moreover all the players were very big boys, most of them owned by major chemical groups. In a sense there were two profit points: in the chemical side of the group making the polymer; and in the fibre division making a profit on the yarn or fibre sales. Well making profit on yarn and fibre in the good old days, but by the seventies profits had become losses and because of the scale of the industry, they were in some cases huge losses, many millions of pounds. But because of the industry structure-with the textile fibre business area of a group being protected by a wide spread of chemical sector profits – the normal rules of competition did not apply. No one went bust, at least not in the shorter term.

Inevitably the losses began to change the way the business was being run. Advertising and promotion budgets were drastically reduced and so the brand names became weaker. The focus shifted to fabric development and closer collaboration with influential specifiers at retail level.

This was the death knell of the major branded fibre business which suddenly lost its raison d'etre. It was not a total shut-down of brands, but the branding focus moved to niche sectors such as Lycra for swimwear and Tactel for sport apparel.

In The Last Chapter we looked at the damage done to synthetics by the fashion change which began in the late sixties. In addition to the problems of increasing over-production and growing competition, fibre producers had to contend with a diminishing demand.

In Britain political problems created a more sombre mood, with the three day week, and power cuts affecting consumer purchasing.

Perhaps in theory this is where the 'big engines' offering lowest prices for the generic product should have come into play. However there were other external forces at work.

Over a number of years the policy of the body operating GATT – General Agreement on Trades and Tariffs – had been to reduce the duties in the West on imports from developing, low-cost manufacturing countries as part on international economic development. In such countries the first phase of industrialization was frequently in the textile sector. The abundance of cheap labour in these countries was particularly suited to textiles in that era and this was reflected by the amazingly low prices for garments – not fabrics or intermediates but finished clothing products.

By the mid seventies the latest round of GATT cuts on import duties into European markets was playing havoc with local textile industries. More and more companies were going out of business. Apart from Portugal – and specialist sectors – fabric producers and garment makers in Europe were completely out-priced. Even the emphasis on new technology in weaving and knitting was not enough to offset the power of ultra-cheap labour.

The problem for fibre producers and for fabric producers was that their customers, the garment makers, could not match garment imports. So no matter if fibre and fabric makers were competitive at their own level, it meant nothing if their customers could not compete with very cheap garments. It was once said that even if ICI Fibres gave their fibre free of charge to UK customers they would still be uncompetitive against imports.

So beset by a fashion change which was contrary to their strengths and prices squeezed by over-competition; and with their secondary and tertiary customers swamped by third world imports, the wonderful twenty years of branded synthetic fibres was coming to a close. Up like a space-racing rocket the Man Made Magic came down at almost the same steep angle.

Could It Have Been otherwise? Well in hindsight, perhaps yes. The pure clarity of good old hindsight. Because there was so much profit in the business at the outset it attracted new players anxious to get into such a plum profit opportunity as soon as patents permitted. Perhaps more defensive strategic pricing in the early days would have deterred some of the new entrants. On the other hand for a country to have a synthetic fibre plant was almost a requirement for national pride – like a airline. So politics rather than the economics of investing in a fibre plant would have influenced the decisions. Moreover the initial licensees would have been anxious to recoup the not inconsiderable investment costs in a what was to an extent a speculative industry in the early days.

Another approach could have been to operate a business on an entirely different basis. That is not to be drawn by the appeal of high volume production; easier said than done though, in an age when efficient mass production was a key element in industry and attractive from a political and social aspect in terms of employment opportunity. Big factories, high levels of employment, feeding the new growing consumer demand. Significantly the parent chemical companies were attuned to high volume business, not niche specialist sectors. They were prepared to invest in development products but if they did not show real major scale prospects they were quietly discarded.

Another objective was to maximise profit from a product while the patent protection remained. Understandably the strong drive to achieve the highest prices and to expand the scale of production was logical.

But down the line, as competition developed and a fashion change was influencing the market demand, there was conceivably an opportunity to change the strategy. To develop a plan to move progressively out of basic product and into specialist niches. Yet even if this policy had made sense it would have been a brave or reckless chairman to pull the plug on the core business. That is not to deny that fibre producers were trying to invent their way out of the market stalemate and to stop banging heads like battling elephants. Some of them were trying hard. But the obsession with concentrating on the main high volume fibres tended to put the new niche products on the side-lines. Volume fibres were 'here and now' sometimes contributing to cash flow if not to profit. How much investment would have to go into invention and development; and how long to wait for a reasonable payback? Would the market pay the price? A fly on the wall at a high-powered meeting of marketeers and R&D scientists might have heard the comment – from a scientist – 'You can have anything you want, a chameleon fibre?, but...it would be very, very expensive'.

You Could Point To one producer that did adopt this strategy and that it was successful. But more in the long term rather than as making a lot of difference to their contemporary operation.

This was Du Pont the American giant, inventor of original nylon. Their Lycra, Nomex and Kevlar are still selling as branded products. ICI achieved a winner with Crimplene although it was based on a standard filament polyester yarn and was exposed to competition. Tactel was an ICI second generation fibre which worked.

Possibly the answer could have been a better balance between slugging it out in the basics – polyester staple fibre, nylon for hosiery and warp-knitting – and putting reasonable investment into new products to re-shape the structure of the industry re-building the industry for a new future.

Ultimately though with the forces at work it was inevitable that the era of branded synthetics, such as Bri-nylon, Terylene, Dacron, Orlon, Courtelle would have to come to an end. But from the first rapturous reception of nylon stockings in 1940 to the decline and fall some forty years later, the synthetic fibre industry had been something rather special, that unique concoction of science and fashion – Man Made Magic.

Today – and tomorrow?

It depends of course on what quality and make you are wearing but with a man's shirt there is a fair chance that it may include a percentage of unbranded polyester with cotton. There has been a quiet return of interest in 'easy-care' and 'non-iron' as now offered on shirt labels and packaging. However it is 'not the sizzle but the steak' which is being sold; the labelling does not make a feature of the polyester content. You probably have to fish for the care instructions down the seam to see what the fibre content and blend levels are.

Interestingly though, the fabric is often soft with a kind handle. Blend technology and fabric finishing has improved considerably over the years. Just for comparison if you are looking at a Jermyn Street shirt it will almost certainly still be 100% cotton – and proud of it. The half dozen or so makers in Jermyn Street were never seduced by easy care.presumably their original clients back in the twenties and thirties did not have be overly concerned about self laundering. Bertie Wooster had more compelling engagements at the Drones club on his mind. And today despite their much broader customer church they stick to their guns and traditions. Their unique characteristic of a collar which never sits quite right and buckles under the fold causing the tie to sit askew, is a much valued indicator of British good taste – and in politicians good judgement.

In The Current Clothing spectrum there is judicious use made of synthetics. Many fabrics used for skirts, trousers, suits and tops include a percentage of elastane as well as other man-made fibres. Generally it is a minor amount – 2-6%. It does not impart much in the way of stretch, but it does seem to provide 'give' which can be claimed as a comfort factor. It also seems to improve fabric appearance and wrinkle resist performance. One of the few still active fibre brands is Du Pont's 'Lycra' for its elastalane stretch fibre.

A current Japanese fabric uses a combination of 41% acrylic, 36% polyester, 28% viscose and 2% elastane for vests and t-shirts under the name 'heatech'. The fabric is silky and fine. Claims for it include active heat generation, heat retention, odour control, anti-static, shape retention. Because of the wicking effect perspiration dries quickly and the fabric remains dry and refreshing. It does seem to work. Ironically as discussed earlier Bri-nylon shirts always did this if the fabric structure was right.

Synthetics sometimes combined with latest generation of rayon are being used creatively to design fabrics in a wide range of textures and characteristics: linen, silk, exotic animal hairs – almost infinite variety. The odd thing is that the man-made fibres are not necessarily being deployed to exploit their performance properties but because they are good for structuring interesting new fabrics. A pity perhaps that this aspect of synthetics had not been a greater influence back in the desperate days of the late seventies. Maybe a policy which focused on blends with a natural look would have been a better strategy; hindsight is so perceptive. The idea of 100% or maximum synthetic content seemed so right at the time. What was best for the consumer – if only he or she knew it.

Nylon of course has never been challenged in the hosiery business: tights and stockings. Fifteen denier, in old terminology, is still the basic. Special occasions may require 'whisper thin' and these are catered for by much finer yarns down to six denier.

Stocking technology has developed and new structures and yarn combinations – allied to judicious use of electrometric high stretch yarns have resulted in perfectly acceptable self hold-up stocking which some women and lots of men prefer to tights.

We can speculate about whether something will come along to replace nylon for hosiery. It would have to be good to compete with nylon's exceptional modulus of elasticity at a competitive price and in a form which could fit easily into established manufacturing set-ups.

And what could it offer that nylon can't? Even if it

could supply the everlasting pair of tights who would want that? Obviously not manufacturers who certainly want to make more than one sale to a consumer even with the prospect of huge new markets in developing countries. Although at first the idea may seem attractive to consumers – it was always a 'magic match' fantasy – nylon hosiery has now become accepted as a consumer disposable. And apart from ultra-fine luxury stockings one wonders how much appetite there may be for an everlasting pair of tights – at least in the West. But will environmental concerns change this?

Some new drastic technology could come along: extruding a form of polymer – let's fantasise on a modified elastane – straight onto a leg shape former with some clever means of instilling micro-pores. A bit like making a latex condom, except for the micro-pore holes of course. A similar technique could be postulated for body shape clothing.

We haven't said much about non-woven fabrics: fabrics which have not been woven or knitted in a conventional way but are made up of fibres such as polyester bound together in a flat web by adhesives and special finishing.

They were the silent voice of synthetics. They were never actively promoted to consumers because they were essentially utilitarian. But they are now very much part of everyday life; and also play important roles in industry.

One of the brands which had coinage is Terram melded fabric originally developed by ICI Fibres. The technology employed a heterofil yarn. This type of fabric is used in separation processes, filtration and erosion control.

Synthetics Have become the norm for many types of luggage; light and strong. A trip down memory lane recalls one of the earliest promotions for nylon luggage when the British team attending the Commonwealth games in Australia in the 1950s was equipped with smart nylon brown suitcases emblazoned with team symbols and red/white/blue flash courtesy of British Nylon Spinners.

In rainwear nylon has become a basic, especially in lightweight tops and clothes for hiking and golf.

The final sector in which synthetics have not only re-invented themselves but have virtually created a new industry, is in in Active Sportswear and perhaps not so active sportswear. This term covers a wide field from fabrics designed to keep the sportsman cool in sports such as athletics, tennis, football, cycling and rowing to fabrics aimed at maintaining body heat in ultra-cold conditions.

The principle involved in cool comfort is not to help the fibre absorb perspiration, soaking up the sweat like cotton, but to transport the moisture to the surface of the fabric and disperse it by evaporation. And it seems that this principle, plus advances in micro-fibre technology, is now being used for the latest active sport fabrics; ultra-light and silky. Rugby players look much sleeker these days not only because of less bar time and more gym time. It's more difficult to hang onto a slippery scrum-half too.

Swimmers have also benefited considerably, so much so that regulations are being discussed to ban superfit one-piece racing suits in the Olympics.

In one form or another synthetic fibres and some

cellulosics, are still alive and contributing to many aspects of modern life.

Today we have a scene in which man-made fibres are established in two diverse ways: they have become integrated into the general range of fibres available for textile use; and also in highly developed forms are fitting into specialist niche sectors.

In the field of contemporary haute -couture there are reports of the Japanese designer Issey Miyake using new technologies with polyester to create original styles in pleated clothing. As well as representing unique fashion design these clothes are light, hard -wearing, have high resistance to wrinkling and are washable. And it seems that polyester is being used because it is the best fibre for the job. Somewhay different from the early use of synthetic fibres in high-fashion where they were under sufferance compared with pure silks or fine woollens . The same designer is repeatedly developing advanced garement making techniques using re-cycled polyester from PET bottlesor fabric waste. Perhaps a new direction for synthetic based apparel fabrics?

But where do we go from here? New superdooper yarns and fibres or totally new textile technology which does not require yarns and fibres as we know them?

We have already looked at the possibilities for stockings and tights and concluded that nothing comes to mind in terms of traditional hosiery manufacturing.

In the medium term that is probably a fair assumption as well for most other apparel types.

One area of opportunity might be in extending the range of truly washable fabrics and garments, all too often one sees the firm direction on some fine articles of clothing 'Dry clean Only'. Dry cleaning is not perhaps the pleasantest activity and somehow does not seem to be cleaning in a traditional sense – at least not in the same way as good old soap – or perhaps detergent and water. Some people may feel that it takes body from the fabric and that dry-cleaned garments wear-out more quickly. To be fair the dry-cleaning is an established and professional industry and by the numbers of outlets provides good consumer service. But perhaps changing environmental attitudes may mean that home and hotel washable garments will appeal more and more to consumers. A chance then for new fibre and textile technology?

Getting in the futuristic slot we could think about potential global influences which may radically affect people's lives. Climate change, either way: big burnout or new ice-age. Increased long-haul international travel with tougher restrictions on luggage may also provide opportunities for a revolutionary form of clothing.

There may be a requirement for astronaut style pressure suits if we finish up being fired to Australia in capsules like rocket ships. Well it certainly would save some time.

Could the expansion of computer science mean that we shall spend our lives in 'Virtual' hibernation: working, travelling, doing all sorts of things we can't even imagine, in a cyber world...never actually leaving our little terrorist-proof urban hideaways.

Would that mean we shall need some form of apparel not yet thought of even by Dr Spock in Star Trek.

Is it possible that the urban terrorist threat may grow to such shocking proportions that we shall all need protective clothing if we dare to venture out: feather weight flak jackets and trousers and headgear, all looking chic but able to stop bomb fragments and bullets at close range. Add to that list ingenious eardrum protection.

Fantastical perhaps but you never know and when you consider what has happened over the past hundred years...well.

If we have to hazard a guess it seems most likely that major short term developments will be in performance materials. New generations of carbon fibre, Kevlar and Nomex type products replacing steel and other metals. As we have seen from spectacular and terrifying Formula One accidents a passenger compartment in road cars based on the F1 car's safety cell, could potentially be much safer- and that is just one idea for creatively engineered new materials.

Whatever the future for synthetic fibres may be just one thing is certain. It will never be as much fun as in the beautiful, dizzy, days of the sixties when man-made fibres were indeed Magical. God bless them.

some textile terms and definitions

A key source of information about all aspects of textiles and the textile industry is The Textile Institute, 5thfloor, St. Jame's Buildings, Oxford Street, Manchester, MI 6FQ

Acrylic fibre
Synthetic fibre based on acrylinitrile

Alginate
Special effect fibre made from sea-wood

Anti-static
Treatment to dispel static electricity

Bleaching
Treatment to whiten fabric or yarns

Bulked yarns
Yarns treated physically or chemically to create volume or bulk

Casein
Basis for protein fibre

Cellulose acetate
Type of rayon

Circular/weft knitting
Type of fabric production used for stockings, tights, socks and jersey

Cotton
Natural cellulosic fibre in many varieties

Course
Row of stitches/loops across knitted fabric

Double jersey

Knitted fabric in rib or interlock stitch

Elastomer/Elastane
Polymer with high stretchability and recovery

False-twist
Method of using heat-setting property of synthetics to produce stretch,crimp and bulked yarns

Full fashioned
Flat frame knitting: fabric panels knitted into shape (knitwear, stockings)

Flax
Natural fibre used for Linen manufacture

Gauge
Used to indicate stitch content in knitting machine (28 gauge fabric is lighter ,finer than 21 gauge)

Hairs
Natural fibre other than wool and silk

Hollow filament
Man-made continuous filament with central canal

Heat-setting
Using thermoplastic nature of synthetics to set fabric and form garments into permanent shape

Hosiery
Knitted fabric for legs and feet

Jacquard
System for manufacture of patterned woven and knitted fabrics

Linings
Fabric inside jackets ,coats, suits, skirts

Lustre
Light reflected from yarn/fabric surfaces

Matt
Dull effect, de -lustred yarns

Melt spun
Production of synthetic fibres by melting thermoplastic polymer and extruding through Spinnerets

Melt-spun
Including colour pigment(dye) into polymer during melt-spinning process

Pick
Weft thread in weaving across width of loom

Pilling
Small bunches of broken fibre on fabric or garment caused by friction in washing/wear

Polyamide
Polymer used to produce nylon fibre (main types nylon 66 and nylon 6)

Polyester
Polymer used to manufacture textile yarns and fibres (eg Terylene); and other products

Polythene
Polymers based on ethylene/polyolefin for fibres and other products

Polynosic
Form of viscose rayon with improved wet strength and reduced wet swelling

Rayon
First type of artificial silk (made from regenerated cellulose: viscose, acetate)

Regenerated cellulose
Cellulose wood pulp liquefied in chemical bath and extruded as rayon filaments

RH
Relative humidity. %water vapour in atmosphere

Schappe silk
Spun silk made from cut -up raw silk waste

Scouring
Treatment of textiles to remove natural and other impurities

Scroop
Sound when fabric, especially silk,is rubbed together

Shrink-resist
Treatment to make fabrics/yarns reistant to shrinkage in washing

Spandex
Alternative semi-generic name for synthetic stretch yarn (elastane,elastomer)

Spinning
The process used to produce textile yarns

Staple fibre
Man-made fibres of predetermined length

Synthetic fibres
Man-made fibres and yarns made from basic chemical elements

Tricot
Name sometimes used for warp-knitting

Thermoplastic
Can be softened and set in shape by heat

Tyre fabric
Motor tyre re-inforcement

Viscose rayon
'Artificial silk' formed by regeneration of cellulose

Warp
Main line of threads across woven fabric

Worsted
Principal spinning system for wool and blends

Woollen spun
Alternative spinning system for wool produces soft milled cloth

Man made Fibre Trade Marks And Brands

Acrilan
Chemstrand acrylic fibre

Bri-nylon
ICI nylon 66 yarn

Celon
Courtaulds nylon 6 yarn

Courtelle
Courtaulds acrylic fibre

Crimplene
ICI textured polyester yarn

Enkalon
Enka nylon 6 yarn

Nomex
Du Pont nylon derivative for anti-flame fabrics

Lycra
Du Pont elastane stretch yarn

Kevlar
Du Pont nylon derivative for high tensile strength fabric

Tactel
Nylon derived yarn for fabrics to wick away moisture from body

Tencel
Modified viscose with high wet strength, soft handle, anti-wrinkle

Trevira
Hoechst polyester fibre

Author's Background

Ronnie Price started his career in textiles with Courtaulds in Coventry. He did the company's well respected two-year training programme round production sites and textile mills. He attended a special course at Coventry Polytechnic in textile technology and economics.

After a secondment to Ireland he transferred to British Nylon Spinners then to ICI Fibres to set up a Dublin office where he was commercial director.

Returning to the UK he became ICI Fibres export manager and later pan-European menswear marketing manager. During his career he wrote extensively for textile publications.

In the 1960s he completed an international marketing programme at Harvard Business School; and in the 70s studied at INSEAD near Paris. He was awarded in 2002 the Annual medal of the Chartered Institute of Marketing.

After his textile career he became director-general of the Portuguese chamber of commerce; and in 2005 was made Knight Commander of the Order of Henry the Navigator by the President of Portugal for his role in bilateral Portugal UK trade.

Previous publications are two thrillers set against the background of Formula One; and a non-fiction story of the win at Le Mans by a small British Morgan sportscar.

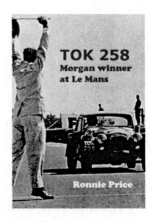

'TOK 258 - Morgan Winner at Le Mans'

'Chicane' and sequel 'On Slicks'

Also from MX Publishing

Rugby Football During the Nineteenth Century

Paul R Spiring

More history and business books at

www.mxpublishing.co.uk

Lightning Source UK Ltd.
Milton Keynes UK
25 November 2010

163440UK00001B/41/P